Tradition and Science of Persian Ink Making

Sadra Zekrgoo

Tradition and Science of Persian Ink Making

Ingredients and Recipes

Springer

Sadra Zekrgoo
Grimwade Centre for Cultural
Materials Conservation
The University of Melbourne
Parkville, VIC, Australia

ISBN 978-3-031-52070-9 ISBN 978-3-031-52071-6 (eBook)
https://doi.org/10.1007/978-3-031-52071-6

This Springer imprint is published by the registered company Springer Nature Switzerland AG
The registered company address is: Gewerbestrasse 11, 6330 Cham, Switzerland

Paper in this product is recyclable.

I dedicate this book to my parents and first teachers, Prof. Dr. Mandana Barkeshli and Prof. Dr. Amir H. Zekrgoo. Throughout my life and academic career, their unwavering support has been the cornerstone of my journey.

Acknowledgments

This book represents the culmination of over a decade of research into Persian manuscript ink recipes, made possible by the invaluable support of the Mary Lugton Post-doctoral Fellowship grant awarded to me in 2021 by the University of Melbourne. I extend my sincere appreciation for the opportunity to work on and complete this volume during my post-doctorate tenure at the Grimwade Centre for Cultural Materials Conservation from 2022 to 2023. The forthcoming second volume, focused on the reconstruction and analysis of inks, is a testament to the continued impact of this grant. The second volume is planned to be published by 2025–2026.

I express my deepest gratitude to my supervisors during my master's degree at Northumbria University and my doctorate degree at the University of Melbourne, where this research originated and expanded. Dr. Jane Colbourne, Associate Professor Petronella Nel, and Prof. Robyn Sloggett shared their extensive knowledge and provided invaluable guidance throughout my studies and beyond.

A special acknowledgment is extended to master calligrapher Keikhosrow Khoroush, whose early teachings on ink construction ignited a passion that has fueled this research for over a decade.

Lastly, I would like to thank Dr. Fariba Afkari for her invaluable assistance in locating treatises on ink making and providing digitized copies of such manuscripts housed at the Tehran University Central Library.

List of Figures

List of Tables

Chapter 1
Script, Calligraphy, and Poetry

1.1 Introduction

The advent of ink has had a profound impact on language, culture, religion, and civilization. Ink with a delivery mechanism, such as quill or pen made possible for speech to be transferred onto supports such as paper, parchment, or papyrus. Carbon ink and iron-gall ink have been extensively researched as historically prevalent writing inks (Phillip 1596, p. 79; Mitchell and Hepworth 1904; Carvalho 1904; Neevel 1995; Thompson and Lindblad 1996; Krekel 1999; Swider et al. 2003; Hahn et al. 2008; Colbourne 2000; da Costa et al. 2013).

However, there are exceptions where the primary writing matter used in certain regions or periods were different than the aforementioned inks such as the formulations used by ancient Persian master calligraphers, where the preferred black writing matter was a carbon-tannate based ink prepared by mixing the two above-stated inks in an appropriate ratio or measurements, and the addition of other secondary ingredients to change the characteristic of the ink based on the calligrapher's preference. These secondary ingredients can affect shade, glossiness, viscosity, fluidity, strength, and longevity of the writing matter.

Despite the considerable research on carbon ink and iron-gall ink, the knowledge of Persian inks is limited, and available information is predominantly in the Persian language (Farsi), thus restricting its international dissemination. Lack of such information, confines the preservation and conservation treatment of manuscripts, works of art on paper, and cultural material where such inks were used on. Understanding the constituents, physical and chemical structures of these inks, as well as the tools and materials employed in their construction, can provide valuable insights to conservators regarding their potential impact on manuscripts' longevity, preservation, storage, and treatment options.

Moreover, investigating the ingredients of Persian inks, whether plant based, animal based, or mineral based, can shed light on the natural resources available in

S. Zekrgoo, *Tradition and Science of Persian Ink Making*,
https://doi.org/10.1007/978-3-031-52071-6_1

Persia at a particular time period, the prevalence of trade within the region, sourcing of materials from other nations, and the regions known for their high-quality materials (such as Indian indigo, Aleppo gallnuts, Cyprus vitriol).

Paper-based objects, such as books, manuscripts, and watercolours, are susceptible to various forms of damage, including physical, chemical, and biological. The ability to identify different inks without intrusive methods can significantly benefit conservators, curators, librarians, historians, and the public with an interest in accessing such artifacts.

This book consolidates over a decade of research on traditional manuscript inks used in Persia and covers the following key aspects:

- Discovering primary sources related to the preparation of Persian inks.
- Translating and categorising ink recipes from classic Persian to modern English.
- Identifying and procuring ingredients required to reconstruct the inks.
- Reconstructing inks using recipes as guidelines.

The process of reconstructing inks helps researchers determine the availability and use of historical ink-making materials and tools. It also provides valuable insights into the intentions of the original ink-makers. Practical experience in ink construction reveals crucial details and gaps in knowledge that might not be evident merely from reading the written recipes.

To facilitate these investigations, the book introduces the history of Arabic script, Islamic calligraphy, and their transformations within the Persian empire and beyond.

There will be a complementary book in the future where a guide to the use of non-destructive and non-invasive analytical techniques to identify and differentiate traditional and reconstructed inks will be covered.

1.2 Islamic Calligraphy and Persia

Before delving into detailed ink investigations, it is essential to grasp the significance of scripture and calligraphy, both aesthetically and sentimentally. Throughout history, religion and art have been means of communication to future generations among living religions worldwide.

1.2.1 Script, Religion and Beauty

Divine words scripted in various languages have symbolized the respective religions in which they were used (Shahi 2015, p. 9). Considerable effort has been put into making script beautiful and elegant, since text is the visual representation of the words of the Supreme Being. As illustrated by this famous Hadith by Muhammad, the prophet of Islam: "God is beautiful, and he loves beauty" (Muslim 2018).

(Shahi 2015) explains that art is creation and artists would consciously or subconsciously have a spiritual connection with The Creator, and the utmost form of creative construction throughout history has been religious. Script and religion have always been closely connected with each other. Within Islam, there is the following verse in the Quran: 'Read (Proclaim!) In the Name of your Lord who created. Created man, out of a clot (of congealed blood). Read (Proclaim), and your Lord is the Most Generous, who taught by the Pen. Taught man that which he knew not' (*Qur'an*).

The term "Islamic art" indeed carries a dual connotation, contributing to occasional confusion. The term encompasses both sacred and non-sacred forms of artistic expression (Zekrgoo 2007).

- Sacred Islamic Art: This category refers to art that emerges as a result of expressions of devotion, reverence, and love for the divine, specifically towards God. It includes works inspired by religious themes, such as calligraphy, mosque architecture, Quranic manuscript illumination, and other forms of art created to honor and express the sublime nature of the divine.
- Non-sacred Islamic Art: On the other hand, non-sacred Islamic art may have originated from a sacred source but does not necessarily maintain a sacred or religious status. These artworks might have had religious themes at their inception but have evolved over time to encompass a broader range of subjects, including daily life, nature, or abstract forms.

The term "Islamic art" is thus inclusive of both sacred and non-sacred forms, highlighting the diverse and rich artistic traditions that have developed within the Islamic world. This distinction is crucial for understanding the breadth and depth of artistic expression within Islamic cultures and civilizations.

For example, A verse from the Holy Quran is inherently sacred due to its religious and holy nature. The reverence for the Quranic verses goes beyond the script itself, encompassing the divine content and significance of the words. On the other hand, a couplet of poetry, despite being in the same style and language, may not inherently carry the sacredness associated with Quranic verses. However, the utilization of the Arabic script in presenting the couplet links it to the broader category of Islamic art, highlighting the significance of the script itself.

The recognition that the Arabic script, irrespective of the content it conveys, is considered Islamic underscores the cultural and religious importance attached to the script. This perception aligns with the view, emerging from the fourteenth century onwards, that writing itself became regarded as a sacred and religious practice. (Soucek 1979, p. 7). According to Safwat, Fehérvári and Zakariya (1997, p. 14), 'The sacredness of the Arabic language played a central role in the development of Arabic script and writing came to be considered as possessing divine power'.

1.2.2 Development of the Arabic Script

Arabic script is related to the Nabataean script, a derivative of Aramaic script, and it was initially named *Jazm* (Safadi 1978, p. 7). After the death of Prophet Muhammad, his secretary Zayd ibn Thabit was ordered by the first Caliph of Islam Abu Bakar, to compile in written form the revelations revealed to Muhammad, which had previously been passed down orally and memorized. The 3rd Caliph 'Uthman, codified the Qur'an (Larcher 2010) and copies were sent to the main regions which were then used as the standard manuscripts from which other copies of the Quran were produced (Safadi 1978, p. 17).

Different forms of the *Jazm* script started forming in different regions of the Arabian Peninsula, amongst which was the Kufic script developed in the city of Kufa. This script had specific measurements and proportions for each letter of the alphabet and by the eighth century, it was the official calligraphic style in which Qurans were written (Safadi 1978; Maḥmudi 2009; Sadabadi et al. 2017).

The Kufic script incorporates straight lines and sharp angles and during the process of reforming the script, attention was given to the development of a more curvaceous form of this script. The cursive style of Arabic calligraphy commonly known as *al-Aqlām al-Sitta* in Arabic or *Ṣiṣ Qalam* in Persian (meaning 6 pens) is divided into six main styles: *Thuluth*, *Naskh*, *Muḥaqqaq*, *Rayhāni*, *Riqā'*, and *Tawqi* (Khatibi and Sijelmassi 1996, p. 79; Sadabadi et al. 2017, p. 140)'. Sources such as Safadi (1978, p. 17) and Shahi (2015, p. 96), mention that a Persian calligrapher by the name of Muḥammad Bin 'Ali Fārsi commonly known as Ibn Muqla ' Fārsi (866-930 C), a master calligrapher and mathematician, was the founder of Arabic cursive calligraphy and laid down the rules for the six calligraphic styles.

- *Thuluth* meaning 'one third' is a large-scale Kufic script first developed in the seventh century during the Umayyad caliphate (661-750) and perfected by the ninth century. It was often used for monumental inscriptions, such as in architectural arabesques, titles, inscriptions, colophons and in the headings of chapters of the Quran known as Suras. The letters are often condensed and may cross one another (Hosseini 1997, p. 14).
- *Naskh* is a cursive script, redesigned by Ibn Muqla' in the seventh century (Hosseini 1997, p. 12) and perfected by other calligraphers especially Ibn al-Bawwāb (d. 1022). More Qur'ans are written in this style than by any other calligraphy style. It was favoured by many (especially the general public) due to its elegance and simplicity, making it easy to read and write (Safadi 1978, p. 62; Maḥmudi 2009, p. 202). It was also the main calligraphic style used during the Abbasid reign (750-1258 CE) for transcripts and texts translated to Arabic from Persian, Indian, Greek and Syriac languages (Khatibi and Sijelmassi 1996, p. 79). Besides the mentioned usage, this style of calligraphy was also used in mosques and minarets, writing Hadith (sayings of the prophet), prayer books, as well as historical and scientific books.
- *Muḥaqqaq* translates to 'meticulously produced' and was another appealing calligraphic style perfected by Ibn Muqla' and Ibn al-Bawwāb (Sadabadi et al.

2017, p. 78). According to Khatibi and Sijelmassi (1995, p. 79), *Naskh* highly influenced both *Muḥaqqaq* and the *Rayhāni* styles (discussed next) while Sadabadi et al. (2017, p. 145) consider *Muḥaqqaq* to be the evolved version of *Kufic* calligraphic style which was originally used for writing Qur'ans.

- *Rayhāni*, an extension of the *Muḥaqqaq* calligraphic style, was developed in the ninth century and was influenced by the first three calligraphy styles. It has a deep sublinear area like *Naskh*, flourishes like *Thuluth* and its leftward curve is reminiscent of *Muḥaqqaq*. It was mainly used for writing Qur'ans and prayer books during the Safavid (1501-1736 CE) period (ibid.).
- *Tawqi'* calligraphic style developed by the ninth century and was influenced by *Thuluth*. *Tawqi'* was the royal script used for writing titles and names of the Abbasid Caliphs and official records. It is said that the famous master calligrapher Yāqut alMusta'ṣami (d. 1298 CE) wrote Qur'ans with this calligraphic style.
- *Riqā'* which is known as the sister of the *Tawqi'* style of calligraphy has more rounded letters and spaced closer together. It is the preferred style of writing throughout the modern Arab world and according to Hosseini (1997, p. 15), it is less popular among Persian writers. During the Safavid period, this calligraphic style was used for writing *firmāns* (royal commands) and occasionally, for writing book titles.

While the mentioned six styles are the main calligraphic styles used for Arabic and Arabic based scripture, there are many other styles inspired from *al-Aqlām al-Sitta*. See Figs. 1.1 and 1.2; they are all different calligraphic styles of writing 'ن و القلم' which in English translates to 'Nuun*(a letter)—by oath of the pen and by oath of what is written' (Qur'an 68:1).

1.2.3 Spread of Islam and the Arabic Script

Arabic script was not created after the emergence of Islam since an elementary form of this script was already being used in the Arabian Peninsula. With the rise of Islam on the world stage, Arabic script was perfected, and spread via the Islamic empire throughout different parts of the world. As lands were conquered and the religion of Islam spread, some countries adopted the Arabic language as their official language and/or Arabic script became the official script. In other words, some countries surrendered their native languages while other countries kept the native tongues but adopted the script (Fazaeli 1983, p. 118).

(Blair 2006, p. 3) suggests the original Arabic alphabet was derived from Aramaic script. Since then, it has developed to its current form, and was adapted to fit other languages, starting with new Persian and Turkish; Dravidian Moplah (Indian dialect) Malagasy spoken in Madagascar; and Malay, the language spoken in Malaysia. Arabic alphabets were used by the Turks up until 1929 when they switched to Roman alphabets (Sakkal 2016).

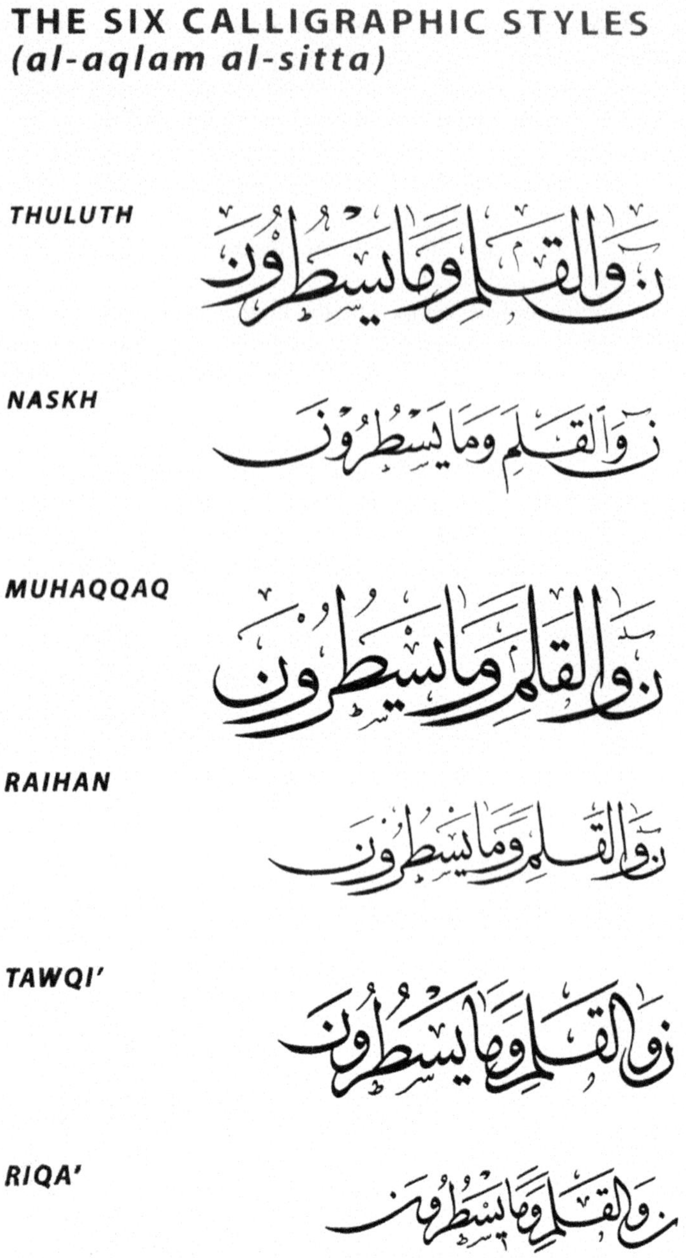

Fig. 1.1 The six primary calligraphic styles (Shish Qalam/ Al-Aqlam alSittah)– Picture taken from an exhibition in Islamic Arts Museum Malaysia (IAMM) 2011. Edited by Zekrgoo 2017

Fig. 1.2 Other calligraphic styles of the Arabic script– Picture taken from an exhibition in Islamic Arts Museum Malaysia (IAMM) 2011. Edited by Zekrgoo 2017

Arabic writing first entered Iraq (Tan et al. 1999, p. 41) and according to Fazaeli (1983) Egypt was the first region outside of the Arabian Peninsula (West of Arabia) to adopt both the language and script. At the time, Egyptians used the Coptic language and script derived from Greek to write and communicate. In the year 706 CE, the Umayyad Caliph Walid bin Abdul Malik made the Arabic language and script, the official language of the country and by the 11th Century Hijri (1592-1688 CE), the Coptic language and script were completely replaced. Similar effects occurred in Syria, which at the time had many scripts, including the official script at the time, Roman and Greek as well as the Syriac, Sumerian and Aramaic languages used amongst the masses.

As the Arabs continued to conquer northern Africa via countries like Tunisia, Algeria and Morocco by way of Egypt, the religion of Islam continued to spread and so did the Arabic script. The calligraphic style of the Arabic script used was predominantly *Maghribi Kufic* or Eastern *Kufic*, which later made its way to Spain and by the 7th Century Hijri (1204-1300 CE) entered Sudan and Timbuktu (ibid).

After conquering North Africa, Arab armies entered Europe via Spain and Andalusia and ruled this region for a period close to eight centuries. Fazaeli (1983) argues that the Arabic script entered Europe via a secondary path as well, by way of Turkey and the Ottoman empire after the conquering of Constantinople (current day Istanbul) in the 15th Century. By the seventeenth century and with the spread of the Ottoman empire, some Greek cities and Balkan countries adopted Arabic script as well.

As the religion of Islam spread throughout Persia, it replaced the Avestan and Zoroastrian religion and the *Kufic* calligraphic style of the Arabic script replaced the Sasanid *Pahlavi* script. From the end of first century Hijri (around 700 CE), the Persians adopted this script and eventually added 4 more letters to the 28 lettered Arabic alphabet by adding dots or lines to already familiar letters; پ، چ، ژ، گ. These letters are the English equivalent of P, CH, ZH and G (Tan et al. 1999; Daniels 2014) which derived from letters ک، ح، ر، ب, the English equivalent of K, H, R and B. Unlike English language, there are no connecting letters to form a new sound in the Arabic script. The addition of new letters by the Persians were revolutionary and inspired other languages adopting the script to add letters so that it can fit better with the language. ‘Thus, if Arabic alone had been taken around the world, Arabic script might not have been so readily adapted to so many other languages’ (Daniels 2014, p. 30).

In the first century Hijri (622-718 CE), the Islamic religion entered India when Muhammad Bin Ghasim Bin Hujjaj Bin Yusof conquered the ‘Sindh’ region (current day Pakistan) and Muslim migrants inclusive of Arabs and Persians settled there. Instead of the Sanskrit language and script, both Arabic and Persian became the most common languages of communication (Fazaeli 1983, p. 121).

Afterwards, from the year 1001-1002 CE with the army of Sultan Mahmud Ghaznavi, Islam was once again re-introduced to India and in 23 years (early part of 11th Century CE), Islamic culture and Arabic script combined with Persian traditions became the norm in northern and central India. From 1175-76 CE onwards, the Timurid dynasty of Persia—of Mongolian descent and influenced by Persian culture—made its way to India and created the Mughal dynasty of India. The Persian-Islamic civilization spread throughout the region and Persian became the official language of the court.

Arabic script spread East as far as China most probably via the Persians. Countries that adopted the Arabic language and script or kept their native language but adopted the script throughout different periods of history include the Arabian Peninsula, Iraq, Syria, Lebanon, Jordan, Egypt, Libya, Tunisia, Algeria, Morocco, parts of China and India, Pakistan, Iran, Afghanistan, Indonesia, Malaysia, and parts of the Balkans.

1.2.4 Calligraphy and Poetry

Considering the importance of script, and the value and beauty of calligraphy, it was no surprise that scribes and calligraphers were the first artists who gained personal and biographical fame (Shahi 2015, p. 96) The practice required extensive training, daily practice, and adherence to various rules and regulations, especially when writing Holy Scriptures. For instance, when writing the Torah, the scribe first performs their ablutions (religious ritual cleansing of self), prior to commencing writing. Text was to be copied from a previous scripture and writing from memory was not allowed. With Christianity, monastery nuns oversaw writing scriptures, painting illuminations and binding books. Calligraphers of Holy Scriptures were considered of the highest ranks.

The Abrahamic faiths (i.e., Judaism, Christianity and Islam) teach that man was created in the image of God. Since art is creation, man is considered both a work of art and an artist (ibid, 120). The art of calligraphy, as the mother of all forms of art in Islam, has been highly esteemed, and calligraphers were the first artists to gain fame. Master calligraphers would write treatises on the art of calligraphy which included information on the mathematical proportion of the alphabets, the type of paper to write on, methods of dyeing paper, types of reed pen to use, placement of writing, how to construct different pigments and dyes, and guidelines on preparing different ink types (Mayil-Hirawi 1993; Afshar 2011). Before elaborating on the practicalities of preparing calligraphy materials, the authors would first focus on the manners and the state of mind needed to start walking on the pathway to becoming a good calligrapher, such as: humbleness, cleanliness, patience, and sacrifice to name but a few.

Historically the art of calligraphy passes down a chain of master to pupil (Blair 2006, p. 241). This section focuses on Persian scribes and master calligraphers who passed down treatises on calligraphy and more importantly ink making, the focus of this book. In Persia, several scripts existed before being replaced by the introduction of Arabic script. The pre-Islamic Persian scripts are known as: (1) *Cuneiform*, a wedge-shaped style of writing used in ancient Persia and Mesopotamia; and (2) *Pahlavi*, a script derived from Aramaic (Figs. 1.3 and 1.4).

Since the pre-Islamic days of the region, poetry was considered the highest form of art and poets were highly respected, especially in Persia where poetry is still highly valued today. Strict rules dictate the writing of traditional Persian

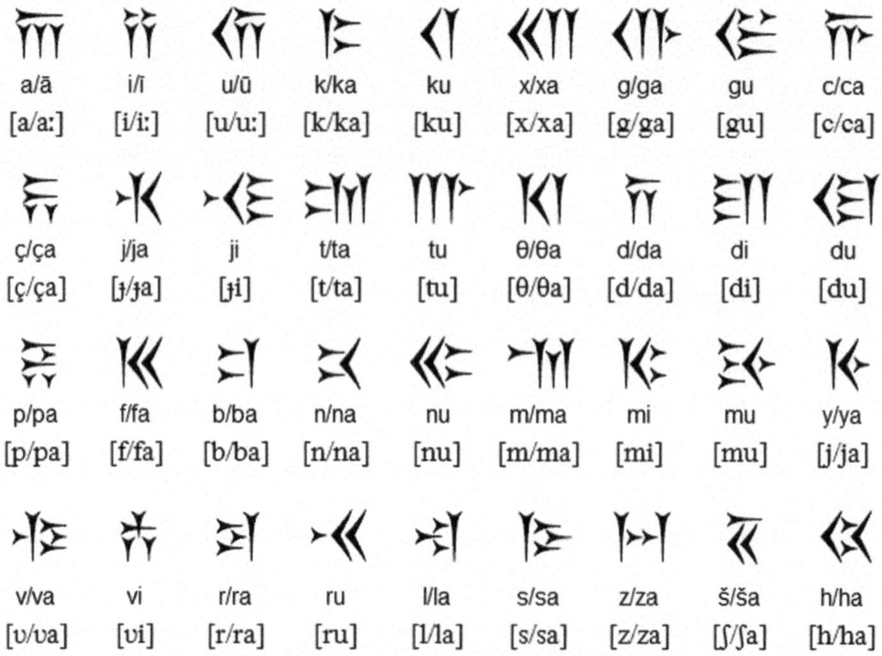

Fig. 1.3 Old Persian Cuneiform. *Source* https://omniglot.com/writing/opcuneiform.htm

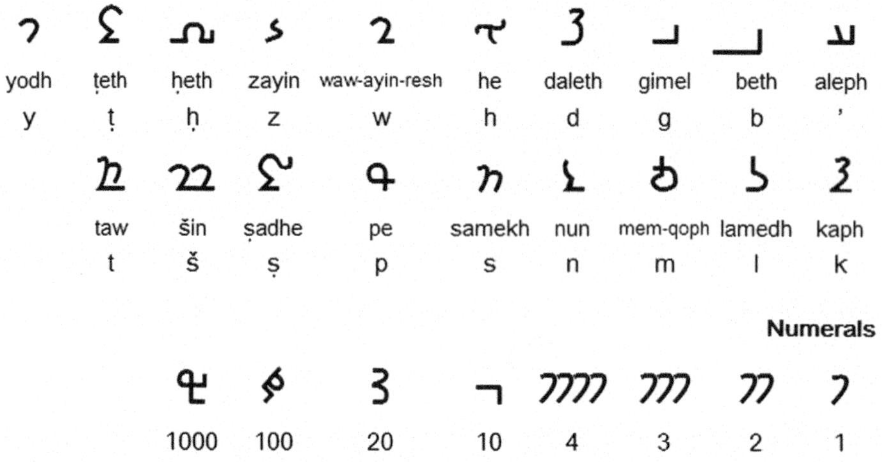

Fig. 1.4 Middle Persian Pahlavi Script. *Source* https://omniglot.com/writing/mpersian.htm

poetry, where rhyming is central to these rules. An example is the poem by Sa 'di (1184-283) translated to English in a similar rhyming fashion to its original Persian form:

بنی آدم اعضای یک پیکرند   که در آفرینش ز یک گوهرند
چو عضوی به درد آورد روزگار   دگر عضوها را نماند قرار
تو کز محنت دیگران بی غمی   نشاید که نامت نهند آدمی

Human beings are members of a whole,   in creation of one essence and soul
If one member is afflicted with pain,   other members uneasy will remain
If you have no sympathy for human pain,   the name of human you cannot retain

Each verse, commonly known as *Bait* in Persian and couplet in English, comprise two smaller sections called *Miṣra* ‘. For this particular poem, the last word of the first *Miṣra* ‘ rhymes with the last word of the second. When writing poetry, there is always a gap between the two *Miṣra* ‘ and once one verse (*Bait*) is completed, the second verse is written beneath the first.

Many of the treatises left by master calligraphers are in the form of poetry. This is due to the cultural importance of poetry, as well as rhymes acting as mnemonics, which aid memorisation. The following is a poem (and its English translation) by fifteenth century master calligrapher Solṭān ‘Ali Mašhadi (15th C) on the topic of the pathway to become a respected calligrapher:

ای که خواهی که خوشنویس شوی   خلق را مونس و انیس شوی
خطه خط مقام خود سازی   عالمی پر زنام خود سازی
ترک آرام و خواب باید کرد   وین زعهد شباب باید کرد
سر به کاغذ چو خامه فرسودن   زین طلب روز و شب نیاسودن
ز آرزوهای خویش بگذشتن   وز ره حرص و آز برگشتن
نیز با نفس بد جدل کردن   نفس بد کیش را زدن گردن
تا بدانی جهاد اصغر چیست   بازگشتن به سوی اکبر چیست
آنچه با خود را نمی داری   هیچ کس را بدان نیازاری
دل میازار گفتمت زنهار   کز دلازار حق بود بیزار
همه وقت اجتناب واجب کن   از دروغ و غیبت و بهتان
ورد خود کن قناعت و طاعت   بی طهارت مباش یک ساعت
حیله و مکر را شعار مکن   صفت ناخوش اختیار مکن
هر که از مکر و حیله و تلبیس   پاک گردید گشت پاک نویس
داند آن کس که آشنای دل است   که صفای خط از صفای دل است
خط نوشتن شعار پاکان است   هرزه گشتن شعار نادان است

Oh, you who want to write beautifully and want to become a companion to the god’s creations. Oh, you who want to be the master of the domain of script and want the world to know your name. You need to refrain from sleep and relaxation and must dedicate your youth to this path. Practice constantly on paper until seeking knowledge keeps you up all day and night. Let go of your dreams and desires and return from the path of greed and lust.

You would also need to battle and defeat your ego until you realize what this small sacrifice is for and what it means to return to the greatest one (the creator). Do not do to others what you would not do to yourself. Remember to never be cruel for justice cannot survive in the hands of the cruel. Always refrain from lying, spreading rumours and vilification. Make devotion (to him) and contentment the pillars of your life; do not go an hour without being cleansed.

Don't make deception and trickery become your chant, refrain from having bad attributes. Whomever gets cleansed from deception, trickery, and guile, becomes a cleansed writer (calligrapher). Only one who is familiar with the heart understands that the purity of the script comes from the purity of the heart. Calligraphy is the slogan of the cleansed while lewdness is the chant of the ignorant.

Chapter 2
Master Calligraphers and Their Treatises

Many Persian master calligraphers contributed to the development and evolution of the Arabic script, most of whom were ink makers and they passed down information on calligraphy and ink making—amongst many other information—by producing treatises. Below are biographies of some of these influential calligraphers, where they resided, and their path of travel.

2.1 Most Influential Master Calligraphers Who Contributed to Ink Making

See Table 2.1.

(i) **Ibn Muqla' Fārsi (886–940 CE/272–328 Hijri) ابن مقله فارسی**

Ali ibn Hasan ibn Abdullah more commonly known as **Ibn Muqla' Fārsi** (886–940 CE) was born in Baghdad (current day Iraq), but his ancestors were hailed from the Fars province of Persia, hence his name (Fazaeli 1983, p. 295). He is the reputed inventor of the six styles of calligraphy as well as the first cursive style of Arabic calligraphy called *Naskh* which he excelled in (Encyclopedia-Britannica 2017). Thackston (2001, p. 25) puts the dating of the creation of these 6 styles to be 922 CE. According to Kuhdezi (1831), he was the first person who added soot to the commonly used metal—tannate Ink. His recipes can be seen in the 15–sixteenth century treatise called *Resāle dar Bayān-e, Kağaḏ, Morakkab, va Hall-e Alvān* (Treatise on paper, ink and creating colours) and other treatises however there hasn't been a treatise found which was left by Ibn Muqla ' himself.

Among his students were Ismail Bin Hammaad Johari Farabi, Muḥammad Bin Ismail Baghdadi, Abu Ishagh Bin Halal Sabi, Shams alMa'ali Qabus Bin Vashmgir, Hasan bin Marzban Seirafi, Ahmad Bin Hussain Alqazari Adib,

S. Zekrgoo, *Tradition and Science of Persian Ink Making*,
https://doi.org/10.1007/978-3-031-52071-6_2

Table 2.1 List of most famous master calligraphers who contributed to Persian ink making and their treatises

Calligrapher/ Writer	**Treatise**	**Dates**
Ibn Muqla' Fārsi ابن مقله فارسی	No known treatises of his own. His recipes on ink making is available in the treatise "Medād al- Ḵoṭuṭ" (Ink for Calligraphy) مداد الخطوط by Mir 'Ali Heravi (16th C-a).	d.940 CE
Ibn Bawwāb ابن بواب	No known treatises of his own.	d.1022 CE
Ḥobeyš Teflisi حبیش تفلیسی	Bayān al- Ṣanā'āt (Articulating Crafts) بیان الصناعات	d. 1203 CE
Yāqut Musta'ṣami یاقوت مستعصمی	No known treatise of his own. Ḥossein'Aqili Rostamdāri (19th CE) in his treatise Kašf al-Ṣanāye' (Discovering Crafts) attributes a recipe to him. کشف الصنایع	d.1298 CE
'Abdollāh Ṣayrafi عبدالله صیرفی	Ādāb-e Ḵaṭ (The Manners of Calligraphy) آداب الخط	d. 1345 CE
Mir 'Ali Tabrizi میر علی تبریزی	No known treatise of his own. His recipes on ink can be found in the treatises Resāle dar Bayān-e Tariqe-ye Sāḵtan-e Morakkab-e Alvān va Kağaḏhā-ye Alvān (Treatises on methods of creating coloured inks and papers) by Anonymous (16th C) and "Medād al-Ḵoṭuṭ" (Ink for Calligraphy) by Mir 'Ali Heravi (d. 1544 CE). رساله در بیان تریقه ساختن مرکب الوان و کاغذهای الوان مداد الخطوط	d. 1446 CE
Maḥmud Šāh Bahmani محمود شاه بهمنی	Resāle-ye Buqalamun (The Turkey Treatise) This is the Persian translation of sections of the Arabic treatise Umdat al-Kuttāb by Ibn Bādis Ṣunhāji (C. 1007-1062) رساله بوقلمون (عمدت الکتب)	d. 1518
Solṭān 'Ali Mašhadi سلطان علی مشهدی	Ṣerāṭ al-Soṭur (The Bridge of Lines) صراط السطور	d. 1520 CE
Mir 'Ali Heravi میر علی هروی	Medād al-Ḵoṭuṭ (Ink for Calligraphy) مداد الخطوط	d. 1544 CE
Majnun Rafiqi Heravi مجنون رفیقی هروی	• Rasm al-Ḵaṭ (Rules of Calligraphy) رسم الخط • Savād al-Ḵaṭ (Knowledge of Calligraphy) سواد الخط • Ādāb al-Mašq (The Manners of Writing) آداب المشق	• d. 1544 CE
Bābā Šāh Iṣfahāni باباشاه اصفهانی	Ādāb al-Mašq (The Manners of Writing) آداب المشق	d. 1588 CE
Qāżi Aḥmad Qomi قاضی احمد قمی	Golestān-e Honar (Rose Garden of Art) گلستان هنر	1519-1582 CE
Fatḥollāh Sabzevāri فتحالله سبزواری	Osul va Qavāed-e Ḵoṭuṭ-e Sette (Doctrines and Rules of The Six Styles of Calligraphy) اصول و قواعد خطوط سته	d. After 1648 CE
Ḥossein 'Aqili Rostamdāri حسین عقیلی رستمداری	Ḵaṭ va Morakkab (Calligraphy and Ink) خط و مرکب	d. After 1571 CE
'Ali Ḥosseini علی حسینی	Kašf al-Ṣanāye'(Discovering Crafts) (Ḥosseini, 19th C) کشف الصنایع	19th CE

Hasan Bin Ali Ma'ruf Banahuj, Ali Bin Yusof Qafti, Ibn Kamunah Muḥammad Samsamani and Muḥammad Bin Asad, amongst others (Fazaeli 1983, pp. 295–300).

During his lifetime, Ibn Muqla' served as a *Vizier* (minister) to three Abbasid Caliphs (Rulers). Unfortunately, there is not even a single page of manuscript that is in this calligrapher's handwriting and all the ones attributed to him are forgeries (Khatibi and Sijelmassi 1996, p. 132). After Ibn Muqla 's death, his daughter carried on the teachings of the father to Ali Bin Hilāl (11th C) commonly known as Ibn al-Bawwāb (d. 1022) who continued on perfecting Ibn Muqla's six style o calligraphy (Shahi 2015, p. 97).

(ii) **Ibn Bawwāb (950/1000–1022 CE/d. 413 Hijri) ابن بواب**

After Ibn Muqla' the most famous calligrapher is Abu alHasan Ali bin Halal better known as Ibn Baww*ā*b (d. 1022). Fazaeli (1983, p. 300) explains Ibn al-Bawwab was born most probably in the late tenth century during the Buyid Dynasty of Persia (934–1062 CE) in Baghdad. He came from a poor family and his father was a doorman, which is why he was commonly known as Ibn al-Baww*ā*b, meaning 'son of doorman'. He undertook apprenticeships with many master calligraphers including Muḥammad Samsamani and Muḥammad ibn Asad, who were students of the great Ibn Muqla'. He was famous for his cursive scripts *Rayhāni* and *Muḥaqqaq* as well as *Naskh*. Ibn Bawwab also perfected the 6 calligraphic styles invented a century earlier by Ibn Muqla'. It is said he wrote over 60 Qur'ans. (Roxburgh 2003) informs that he excelled in the art of calligraphy to the point that he was given the position of librarian to the sultan al-Dawla (988–1022 CE).

(iii) **Ḥobeyš Teflisi (twelfth century/sixth century Hijri) حبيش تفليسى**

Ḥobeyš Bin Ebrāhim Bin Mohammad Teflisi was a Persian writer, healer, and astrologer of the twelfth century. He was either born in Tbilisi (Georgia) or Ghazni (Afghanistan), and he was known to have lived in Konya (Turkey) for long periods of his life (Afshar 1957, p. 282). He has left multiple treatises on the subjects of medicine, astrology, Qur'anic commentaries, artist materials (inks and pigments), and Persian and Arabic literature (ibid.)

(i) **Yāqut Musta'ṣami (thirteenth century/seventh century Hijri) ياقوت مستعصمى**

Another well-known calligrapher who contributed to the perfection of six style of calligraphy was Yāqut Musta 'ṣami (d.1298) and his six styles of writing is known as the Yāquti style. According to (Soucek 1979, p. 8), his work was imitated extensively in 14–15th CE Iran. Jamāl al-Ddin Yāghut al Musta 'sami (d. 1298 Baghdad) was a servant of the final Calif of the Abbasid Dynesty, al-Musta 'sam-Billāh Abu-Ahmad Abdullāh bin al-Mustansir-Billāh which is why he referred to himself as Yāqut Musta 'sami (Yāqut of Musta'sem) (Fazaeli 1983, p. 313). The period after the time of Yağut is knowns as the highlight of calligraphy.

He had many students and followers, but his most prized apprentices were Arqun bin Abdullāh Kāmeli, Nasrullāh Ṭabib, Yusof Mašhadi, Mobarakshāh Bin Qutb Tabrizi known as Zarrin Qalam (golden scribe), Seyed Ḥeydar Gonde Nevis and Aḥmad ibn-e Sohravardi known as Šeyḵzāde (Thackston 2001, p. 8). Each student of Musta 'sami took apprentices of their own.

In the treatises researched, his name is repeatedly referred to, as well as the name Khājah Jamāl al-Din Yāqut, who could be the same person (Islamic-Arts, 2012). It is known however, that he had an apprentice by the name of Siyid Ḥiydar Gundih Nivis (13th C) who mentored, 'Abdollāh Ṣeyrafi bin Maḥmud Tabrizi (13th C) commonly known as 'Abdollāh Ṣayrafi. Yāqut was an expert in Naskh, Tawqi ', Muḥaqqaq and Raihan styles of calligraphy and he developed a new calligraphic style known as Yāqut*i*.

(ii) **'Abdollāh Ṣeyrafi (d. 1345–46 CE/746 Hijri) عبدالله صیرفی**

'Abdollāh Ṣayrafi, son of Maḥmud Ṣeyrafi was born in the city of Tabriz which was the capital of Persia during the Safavid dynasty, and he possibly died in Tabriz as well. Mayil-Hirawi (1993, p. 16) mentions 2 treatises which are left by behind by this calligrapher; *Qalamiyye* (Art of the Pen) and *Ādāb-e Ḵaṭ* (The Manners of Calligraphy) (Bayani 1985). Ṣeyrafi studied with Seyed Ḥeydar Gonde Nevis who was a prized student of Yāqut Musta'ṣami (Thackston 2001). Amongst Ṣeyrafi's most famous pupils are Ḵair al-Din Mar'ashi and his nephew Haji Muḥammad Bandgir Tabrizi. According to Soucek (2011), he passed away after 746 Hijri (1345–46 CE). Ṣeyrafi was most famous for his Naskh and Thuluth style of calligraphy.

(iii) **Ḵāje Mir Ali Tabrizi (1360–1446 CE/752–850 Hijri) میر علی تبریزی**

Another calligrapher from the same city was Mii Ali Soltāni, son of Mir Hasan Soltāni more commonly known as Ḵāje Mir Ali Tabrizi. He was born in 1360 in Tabriz (Qalam 2009) and his date of death is set at 1446. He invented the most common Persian script style called *Nastaliq*, by integrating the *Naskh* style of writing (One of the six main calligraphic styles) with the *Ta'liq* (A secondary calligraphic style) (Minorsky 1959). According to the estimation of Fazaeli (1983, p. 448) this style was the predominant form of calligraphy from fourteenth century Persia until today, and from mid-fifteenth century onwards, 75% of all the written material in Persia was in this calligraphic style. Unfortunately, there aren't many manuscripts left that are in his handwriting however his recipes and advice are passed down and documented in the treatises of other master calligraphers.

Minorsky (1959, p. 100) mentions that Tabrizi taught this style to his son Ubaydullāh and he became quite an outstanding master and that the person who made this script superior was Solṭān 'Ali Mašhadi (d.1519). Fazaeli (1983, p. 453) and Thackston (2001, p. 10) both mention that Tabrizi's son was named Abdullah, but some sources have mentioned Ubaydullāh. They further explain that another pupil of Tabrizi besides his son was Mirzā Jafar Tabrizi who taught this style to the famed calligrapher Muhammad Simi Neišāpuri commonly known as Simi Neišāpuri or Simi Neišāburi.

(iv) **Solṭān 'Ali Mašhadi (1437–1520 CE/841–926 Hijri)** سلطان علی مشهدی

Fazaeli believed that no one became more famous during the course of their lifetime than Solṭān 'Ali Mašhadi. He was born in 1437–38 CE in the city of Mashhad,Iran (Mayil-Hirawi 1993, p. 45). His father passed away when he was at the age of 7 at which point, he raised by his mother. It is said that unlike other children, he spent most of his days practicing calligraphy without a master until he was discovered by Mir Moflesi who was a poet and taught him how to write properly.

He excelled in calligraphy and his fame reached Sultan Hussein Mirza Bayqara Gurkani (873–912), the Timurid ruler of Herat (Afghanistan) who requested Solṭān 'Ali to travel to Herat and become a scribe at his library. After the death of the king and being succeeded by Uzbek kings who did not value his work, he travelled back to his city Mashhad and after battling with smallpox, he passed away at the age of 85 in 1520 (Fazaeli 1983). Mašhadi wrote his treatise *Ṣerāṭ al-Soṭur (The Bridge of Lines)* towards the end of his lifetime in 1514–15 CE/920 Hijri.

There were no sources found that mentions him being a pupil of Tabrizi or ever meeting him. What is known however according to Soucek (1979, p. 166), is that the artists trained at the Tabriz Safavid court, influenced and contributed in the development of Mughal style of painting of India during their migration to this country, and amongst them was Solṭān 'Ali Mašhadi (d. 1519) 'a famous calligrapher who had influenced the Mughal writing styles to the point that his calligraphy became the standard of excellence against which all other works would be judged (Allami 1939, p. 108).

Based on Soucek's words, it is possible to assume that at one point, Mašhadi did train in Tabriz and might have met with Tabrizi. What is factual however is during his travels to Herat (Afghanistan) and prior to his migration to India, he mentored Mir 'Ali Heravi (d.1544–45), yet another famous calligrapher (Bayani 1985, pp. 1060–1061). As seen, there seems to be some discrepancies between the history given by Fazaeli (1983) where he mentions that Mašhadi only travelled between Herat and Isfahan (and passed away there), and the history given by Soucek (1987) where Mašhadi travels to India.

(v) **Maḥmud Šāh Bahmani (d. 1518 CE/924 Hijri)** محمود شاه بهمنی

Whilst not much is known of Maḥmud Šāh Bin Mohammad Šāh Bin Ahmad Šāh Al-Wali al-Bahmani, this person contributed to the art of ink making by translating the colour section of the famous Arabic treatise Umdat al-Kuttāb by Ibn Bādis Ṣunhāji (C. 1007–1062) into Persian in the fifteenth century under the name of *Resāle-ye Buqalamun,* or 'The Turkey Treatise' (Kashani 2010).

(vi) Mir 'Ali Heravi (1465–1544 CE/869–951 Hijri) میر علی هروی

Mir Ali Hiravi was born in Herat, Afghanistan and he studied arts in his youth and was a student of Zein al-Ddin Mahmoud who was known to be an apprentice of Solṭān 'Ali Mašhadi (Fazaeli 1983) although Shokrollahi (Shokrollāhi 2015, p. 63) refers to Heravi as a pupil of Solṭān 'Ali Mašhadi himself. Shokrollahi further illuminates that Heravi wrote a treatise entitled *Medād al-Ḵoṭuṭ (Ink for Calligraphy)* in which he writes a commentary of the treatise *Ṣerāṭ al-Soṭur (The Bridge of Lines)* left by Mašhadi. He believed that due to Mašhadi 's treatise being dominated by poetry, while the rhymes help with memorization, the content may be too confusing for inexperienced calligraphers.

Heravi felt that the details and delicate information must be written in a clear succinct manner so that 'the main points' does not escape the reader when being surrounded by rhymes (ibid, p. 64). Hence, while having poetry, his treatise also has normal commentary as opposed to Mašhadi's treatise which is only poetry based.

Mir 'Ali Heravi became amongst the artists of the royal court of Sultan Hussein Mirza Bayqara and eventually was bestowed with the title of Kātib al-Soltāni meaning the Sultan's Scribe. He resided in Herat until the death of the Sultan in 911 Hijri (1505 CE); after which he travelled in between Mashhad and Herat. After the Uzbeks conquered Herat, many artists were captured by Ubaidullah Khan Uzbak, including Heravi, and were forced to migrate to Bukhara (ibid, p. 63). His death is dated as 951 Hijri (1544 CE), 16 years after he first set foot in Bukhara.

His treatise *Medād al-Ḵoṭuṭ (Ink for calligraphy*) has many recipes, a lot of which were passed down by his predecessors. Allami (1939, p. 109) describes him as a master calligrapher who has left many masterpieces and perfected the calligraphic style of Mašhadi, while Nihavandi (1616), refers to him as the *Qiblat al-Kuttāb* or 'Qibla of Scribes' (Soucek 1979, p. 166).

According to Shokrollahi (2014), after Tabriz, Herat was the most important city that contributed to the development of the Nasta'liq calligraphic style, and this is also credited to the Gurkani Sultans especially Sultan Hussain Bayughra and his royal vizier.

(vii) Majnun Rafiqi Heravi (d.1544 or 1603 CE) مجنون رفیقی هروی

Another famous calligrapher from Herat is Solṭān Aḥmad Majnun Rafiqi Heravi (16th C), commonly known as Majnun Rafiqi Heravi who was the son of Kamal al-Ddin Maḥmud Rafiqi. He was a poet as well as a calligrapher and studied with many master calligraphers from a very early age and was known as Majnun meaning crazed (Fazaeli 1983, p. 482). He was also known as *čhap nevis* (left-writer) since he was left-handed. Majnun Rafiqi Heravi had left 3 treatises; *Rasm al-Ḵaṭ (Rules of calligraphy)* published in 1503 CE/909 Hijri, *Savād al-Ḵaṭ (Knowledge of Calligraphy)* after 1524 CE/930 Hijri and *Ādāb al-Mašq* (The Manners of Writing) (Mayil-Hirawi 1993). Not much is known about his year of birth, but his death is dated to be 1544 CE/951 Hijri (Bayani 1985).

(viii) Bābā Šāh Iṣfahāni (d. 1588 CE/996 Hijri) باباشاه اصفهانی

Bābā Šāh Iṣfahāni (16th C) was another famous master calligrapher who was born either in Isfahan or the foothills of Iraq but reached fame in the city of Isfahan and hence, was known as Iṣfahāni. Reed-College (2014) mentions him to be born in Isfahan. According to (Ernst 1992, p. 280), he studied calligraphy at an early age with the famous Mir 'Ali Heravi. Iṣfahāni left a treatise entitled *Ādāb al-Mašq* (The Manners of Writing) before he passed away in Baghdad (Iraq) either in 1587 or 1588.

Unfortunately, many manuscripts handwritten by Iṣfahāni including *Ādāb al-Mašq* were falsely credited to the work of Mir 'Emād Qazvini (d. 1615), another master calligrapher of the same era famous for his Nast'aliq style of calligraphy (Ghilichkhani 2012, p. 7). According to Fazaeli (1983, p. 492) experts in Nast'aliq calligraphy believe that no calligrapher (classic or contemporary) has superseded the style of Bābā Šāh Iṣfahāni.

(ix) Qāżi Aḥmad Qomi (1519–1582 CE) قاضی احمد قمی

Qāżi Aḥmad ibn Mir-Monshi al-Hoseyni (1519–1582) came from the city of Qom, located south of Tehran, Iran, which is why he is known as Qāżi Aḥmad Qomi (Minorsky 1959). He was the son of the Safavid court secretary Sharaf Al-Din Hossein, commonly known as Mir-Monshi, and from the age of eleven he studied calligrapher under master calligraphers such as including Shah Mahmud Neishaburi, Malek Daylami, and 'Ali Asghar (Farhad and Simpson 1993).

Qomi's most influential work is the treatise *Golestān-e Honar* (Rose Garden of Art) where he writes about the art of calligraphy, paper dyeing, pigments, and inks. This is one of the few treatises that has been translated to English thanks to Minorsky, and has been constantly referred to by scholars.

(x) Fatḥollāh Sabzevāri (16th CE) فتحالله سبزواری

Fatḥollāh bin Aḥmad, bin Maḥmud Sabzevāri (16th CE) more commonly known as Fatḥollāh Sabzevāri. Sabzevāri was born either in Tabriz or in Sabzevar, both of which are in Iran. According to Mayil-Hirawi (1994, p. 53–54), he was the scribe of Shah Isma'il Safavi (1501–1524 CE) at the Tabriz court before migrating to Istanbul and becoming a scribe at the court of Sultan Salim I. his treatise is titled *Osul va Qavā 'ed-e Ḵoṭuṭ-e Sette* (Doctrines and Rules of The Six Styles of Calligraphy) written in 1548 CE/955 Hijri (Neishaburi 2004).

(xi) Ḥossein'Aqili Rostamdāri (16th CE) حسین عقیلی رستمداری

Ḥossein 'Aqili Rostamdāri (16th CE) was a calligrapher (famous for his Nast'aliq stlyle) and scientist who lived in the city of Qazvin (Iran) during the reign of Shah Tahmasp the first (1524–1576 CE/930–984 Hijri) and he has one treatise by the name of *Ḵaṭ va Morakkab* (Calligraphy and Ink) written in 1571 CE/979 Hijri (Qarashi 1996, p. 27). According to Mayil-Hirawi (1994, p. 72) he most probably left Qazvin for Iraq during towards the end of his life.

(xii) **'Ali Ḥosseini (19th CE) علی حسینی**

'Ali Ḥosseini (18–19th CE) with his treatise *Kašf al-Ṣanāye* ' (Ink Making and Book Binding) is the only one that contains recipes on English and European ink in addition to the authentic Persian ones. This is due to the treatise being more recent, and availability of broader information (ibid).

In addition to calligraphers and their treatises mentioned above, there are also treatises on ink making and calligraphy where the authors or scribes are unknown and have invaluable information:

- Anonymous (15th C)—*Resāle dar Bayān-e, Kāğaḏ, Morakkab, va Hall-e Alvān* (Treatise on Paper, Ink and Creating Colours). The content of this treatises is strikingly similar to the treatise of Simi Neišāpuri researched by Porter (1985) and Thackston (2001).
- Anonymous (15th C)—*Resāle dar Bayān-e Ḵaṭ va Morakkab va Kāğaḏ va Sāḵtan-e Ranghā (Treatise on Calligraphy, Ink, Paper and Creating Colours.).*
- Anonymous (16th C) *Resāle dar Bayān-e Tariqe-ye Sāḵtan-e Morakkab-e Alvān va Kāğaḏhā-ye Alvān* (Treatise on Methods of Creating Coloured Paper and Ink).
- Anonymous (16th C)—*Majmu'al-Ṣanāye'* (Collected Crafts).

As mentioned earlier, information on calligraphy was passed down from master to pupil by leaving behind manuscript and treatises. In addition, the calligraphers made their own ink, the recipes of which were also passed down to their pupils and in the same method that the following generation perfected the calligraphic styles, they also perfected the inks.

2.2 Art and Calligraphy Centres During the Safavid Period

Most of these treatises mentioned were gathered during the Safavid dynasty (1501–1736) which was a very important period in art history of Persia. According to Blair (2006, p. 417), it was during this period where the political and cultural unity of the Islamic world slowly started to disintegrate, and the central Islamic lands were separated into three major Empires; the Ottoman Empire of Turks, Mughal Empire of the Indians, and the Safavid Empire of the Persians.

It was also in the Safavid dynasty where Shi'ism became the state religion of Iran and it was in this period where works of calligraphy and different treatise were collected, thoroughly studied, and perfected. In Blair's own words, '*it was the first time that we can speak of an art history of Iranian calligraphy*'. This era also generated Persian paintings and the integration of images with calligraphy; a practice frowned upon and in many cases prohibited by Muslim clerics outside of Persia; there are still debates amongst clerics even today.

During the Safavid period (1501–1736), there were five main centres of art and calligraphy (Sadabadi, Ja'fari and Shahidabi, 2017) and the Nast'aliq style of calligraphy, which is considered the official calligraphic style of Persia evolved within these centres:

(i) **Tabriz Centre (1501–1548 CE)**

After the Safavid empire came to power in 1501 (CE), Shah Ismail I (1487–1524 CE) shifted the capitol of Persia from Herat, which was the capital during the Timurid Dynasty (1370–1507 CE)—to Tabriz. During this shift, many artists and scientists were transferred to this city as well.

(ii) **Qazvin Centre (1548–1598)**

In 1524 CE, Tahmasp I (1514–1576 CE), son of Ismail I came to power and reigned over the Safavid dynasty until 1576 CE. In 1548, Shah Tahmasp I chose Qazvin as his royal court. During this shift, many artists migrated to Qazvin from Tabriz while others migrated from different cities.

(iii) **Mashhad Centre (1556–1576)**

Towards the end of Shah Tahmasp, Ist reign, many artists gravitated towards the royal court of prince Ibrahim Mirza (1540–1577 CE) in Mashhad. Ibrahim Mirza—the grandson of Ismail I and the nephew of Tahmasp I—was a patron of the arts and between 1556–1576 CE, Mashhad became a centre of the art of book binding (book-making).

(iv) **Shiraz Centre (Established during the Timurid dynasty, continued until the end of 16th C)**

Established during the Timurid dynasty, the city of Shiraz continued to be amongst the main hubs of calligraphy and the art of bookmaking until the end of 16th Century.

(v) **Isfahan Centre (1598–1736)**

In 1598, the capital of the Safavid empire shifted from Qazvin to Isfahan. Abbas I (1571–1629), son of Ismail I supported the arts, especially calligraphy. During this period, calligraphy and painting which used to go hand in hand became separate branches of art and calligraphers and painters started creating individual works (Figs. 2.1 and 2.2).

Fig. 2.1 Main art and calligraphic centres during the Safavid period. Image created by author

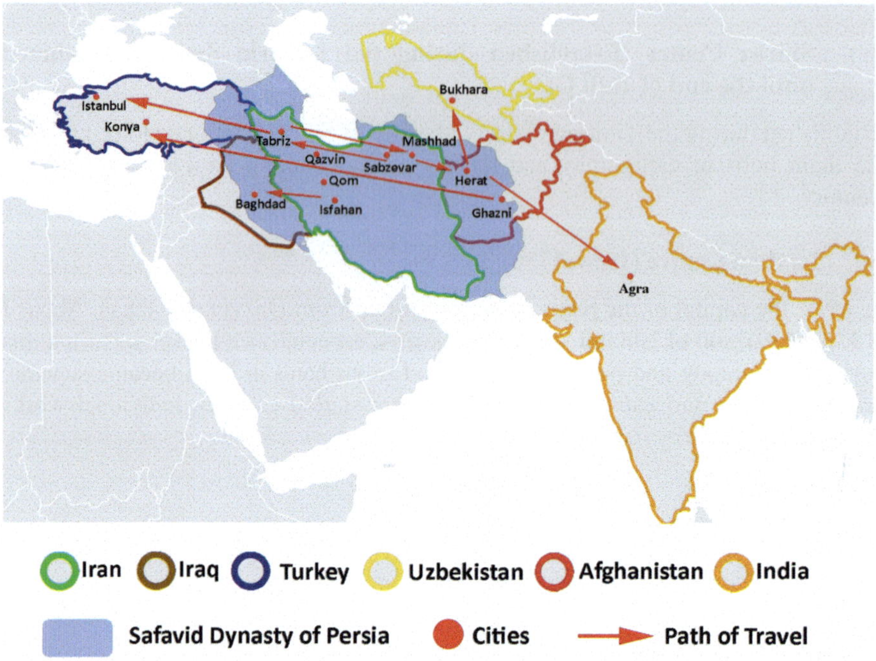

Fig. 2.2 Locations where master calligraphers resided and their path of travel. Image created by author

Chapter 3
Constituents of Persian Inks

Ingredients used in all the different Persian inks mentioned are either plant, protein, mineral, or synthetic based and were divided in to seven main categories (see Table 4.8); colourants, binding mediums, mordants and additives, aromatics, thickening agents, glossing agents, and preservatives. Water is mainly used as the vehicle to make the ink fluid. These categories were created from information gained through interviewing chemists and master calligraphers) investigating the characteristics of each ingredient and discovering their usage in traditional Persian medicine by referring to the plant encyclopedia by Mir-Heidar (2012a–h). Unless specified otherwise, information regarding plants and their medicinal usage, the information has been extracted from the 8-volume encyclopedia.

Calligrapher and ink maker master Khoroush stated during an interview (2015), that traditionally, knowledge of the medicinal qualities of most natural ingredients was broadly practiced and therefore would have informed many of the ink makers in their use. Materials mentioned can belong to more than one category. For instance, a material can be used as a colourant, mordant, and binding medium. Additional factors that can affect the potency and characteristic of ingredients are the geographic location of where a plant grows or where the minerals and ores are found, as well as production and preparation methods. These categories and relevant ingredients (summarised in Table 3.1) will now be outlined in detail.

As mentioned, the ingredients can be further divided in to three main categories of plant, mineral and protein-based materials. Most of the plant-based materials used in ink making were also consumed at the time and to a great degree today, due to their known medical properties they have from a traditional Persian medicine point of view. These properties may have been the primary or secondary reasons behind their use in inks and thus the importance of each plant according to practitioners of traditional Persian medicine will be explained.

Information regarding plant-based ingredients are extracted from the 8-volume book Ma'āref-e Giyahi (Introduction to plants) by Mir-Heidar (2012) and most data relevant to minerals and pigments are obtained from the book Pigment

S. Zekrgoo, *Tradition and Science of Persian Ink Making*,
https://doi.org/10.1007/978-3-031-52071-6_3

Table 3.1 Detailed list of ingredients used in traditional Persian ink making

Category	Common Name	Scientific Name & Formula	Persian Name	Transliteration
Colourant	Lamp black	Carbon (C)	دوده	Dude
	Gall nut	Contains high amount of tannic acid $C_{14}H_{10}O_9$	مازو	Māzu
	Green vitriol	Hydrated iron (II) sulphate $FeSO_4 5H_2O$	زاج سبز	Zāj-e sabz
	Blue vitriol	Hydrated copper (II) sulphate $CuSO_4 \cdot 5H_2O$	زاج کبود	Zāj-e Kabud
	Red vitriol	Potassium ferricyanide $K_3[Fe(CN)_6]$	زاج سرخ	Zāj-e Sork̲
	Yellow vitriol	Hydrated potassium ferrocyanide $K_4[Fe(CN)_6] \cdot 3H_2O$	زاج زرد	Zāj-e Zard
	Marcasite	Iron sulfide FeS_2	مرقشیشا	Marqešišā
	Copper	Cuprum (Cu)	مس	Mis
	Verdigris	Copper Acetate $CuCO_3.Cu(OH)_2$	زنگار	Zangār
	Lapis lazuli	Lapis lazuli $(Na,Ca)_8[(S,Cl,SO_4,OH)_2|(Al_6Si_6O_{24})$	لاجورد	Lājvard
	Green walnut husks	Extracted from Juglans regia (walnut)	پوست گردکان سبز	Pusti Gardakān-e Sabz
	Henna	Lawsonia inermis	حنا	Ḥanā
	Indigo	Indigofera tincotoria	نیل	Nil
	Woad	Isatis tinctoria	وسمه	Vasme
	Myrtle	Myrtus communis	مورد	Murd
	Saffron	Crocus sativus	زعفران	Za'farān
	Madder	Rubia tinctorum	روناس	Runās
	Aloe vera	Aloe vera	صبر	Ṣabr
	Cinnabar	Vermillion	شنجرف	Šanjarf
	Gold	Aurum Au	طلا	Ṭalā
	Silver	Argentum Ag	نقره	Noqre
	Tin	Stannum Sn	قلع	Qal'
	Quicksilver	Mercury Hg	سیماب/ جیوه	Simāb / Jivih
	Brass	Alloy made of copper and zinc	برنج	Berenj
	Red Poppy	From the genus Papaveraceae	شقایق نعمان	Šaqāyeq-e Na'mān
	Red Mallow	Mathiola Incana – From the genus Cruciferae	خیری سرخ	K̲eiri-e Sork̲
	Sumac	Rhus coriaria L - from the genus Anacardiaceae	سماق	Somāq
	Brazilwood	From the genus Caesalpinia	بقم	Baqam
	Logwood	From the genus Leguminopsae	بقم بنفش	Baqam-e Banafsh
	Sappanwood	Biancaea sappan	بقم سرخ	Baqam-e Sork̲
	Indian Catechu	Acacia catechu (L.) Willd - From the genus Mimosaceae	کات هندی	Kāt-e Hendi
	Cochineal	Extracted from insects known as Coccoidea	کوچنیل	Kochenil
	Turmeric	Curcuma domestica val - From the family of Zingiberaceae	زردچوبه	Zard-čube
	Minium/Red Lead	Lead oxide PbO	سرنج	Soranj
	Antimony collyrium	Antimony trisulfide Sb_2S_3	سرمه عملی	Sorme-ye 'Amali
	Orpiment	Arsenic trisulfide As_2S_3	زرنیخ	Zarnik̲
	Asphaltum	By-product of the coal and petroleum	اسفالتم	Āsfāltum

Table 3.1 (continued)

	Pumpkin rind	Rind of pumpkin	پوست کدو	Pust-e Kadu
	Lac	Dyestuff extracted mainly from insects known as Coccoidea	لاک	Lāk
	Chinese Blue / Prussian Blue	Ferrocyanide $[Fe(II)(CN)_6]^{4-}$	چاینیز بلو / پرشن بلو	Chāyniz blu / Pershan blu
	Talc	Magnesium silicate hydroxide $Mg_6Si_8O_{20}(OH)_4$	طلق	Ṭalq
	Lead Ceruse	lead carbonate $PbCO_3$	اسفیداج ارزیز	Esfidāj-e Arziz
	Myrobalan	The fruit of a tree from the genus Combretaceae	هلیله	Halile
	Pigeon's blood		خون کبوتر	Ḵun-e Kabutar
	Bitter orange	Citrus bigaradia L	نارنج	Nārenj
	Urine		آب تاختن	Āb-e Tāḵtan
	Fish oil		روغن ماهی	Roḡan-e Māhi
	Colocynth	Citrullus colocynthis (L.) Schrad	حنظل	Ḥanẓal
Binding medium	Gum Arabic	Extracted from the Acacia Senegal tree	صمغ عربی	Ṣamq-e 'Arabi
	Liquidambar	Extracted from trees of the family Hamamelidaceae	عبیر اشهب / عنبر اشهب	'Abir-e Ašhab / 'Anbari Ashhab
	Sandarac	Extracted from the acacia tree of the genus Tetraclinis articulata	سندروس	Sandarus
	Egg white		سپیده خایه / سفیده تخم مرغ	Sepide-ye Ḵaye / Sefide-ye Toḵm-e Morḡ
	Gum Ammoniac		وشه / وشق	Voše / Vošaq
Thickening agents	Camel thorn	Alhagi maurorum	برگ خار / خار شتر	Bargi Ḵār / Ḵār shutur
	Dog rose extract	Rosa canina	عرق نسترن	'Araq-e Nastaran
	Epithymum	Cuscuta epithymum	افتیمون	Aftimun
	Coral	-	کفک درایِ	Kafak-e daryā
	Gum Arabic	Extracted from the tree Acacia senegal	صمغ عربی	Ṣamq-e Arabi
Aromatics	Damask Rose water	Rosa damascena	گلاب	Golāb
	Chicory extract	Cichorium intybus	عرق کاسنی	'Araq-e kāsni
	Musk	-	مشک	Mushk
	Liquidamber	Extracted from trees of the family Hamamelidaceae	عبیر اشهب / عنبر اشهب	'Abir-e Ašhab / 'Anbari Ashhab
	Dog rose extract	Rosa canina	عرق نسترن	'Araq-e nastaran
Preservatives	Aloe vera	Aloe vera	صبر	Ṣabr
	Salt	Sodium chloride NaCl	نمک	Namak
	Ox-gall	-	زهره گاو	Zahri Gāv
	Henna	Lawsonia inermis	حنا	Ḥanā
Glossing agents	Sugar	-	شکر	Šekar
	Honey	-	عسل	'Asal
	Egyptian sugar candy	-	نبات مصری	Nabāti Meṣri
	Gold	Aurum Au	طلا	Ṭalā
	Silver	Argentum Ag	نقره	Nughrih

Table 3.1 (continued)

Mordents & additives	Alum	Hydrated potassium aluminium sulphate $AlK(SO_4)_2.12H_2O$	زمه	Zame
	Sal ammoniac	Ammonium Chloride NH_4Cl	نوشادور	Nushādor
	Green vitriol	Hydrated iron (II) sulphate $FeSO_4{\cdot}5H_2O$	زاج سبز	Zāj-e Sabz
	Blue vitriol	Hydrated copper (II) sulphate $CuSO_4{\cdot}5H_2O$	زاج کبود	Zāj-e Kabud
	Marcasite	Iron sulfide FeS_2	مرقشیشا	Marqešišā

Compendium by Eastaugh (2008). It should be noted that ingredients which are from the 19th C European recipes translated to Persian by Hosseini are omitted from the list and the focus is strictly on ingredients found in **Persian** ink recipes.

3.1 Colourants

Colourants provide colour via the use of pigments, dyes and chemical reactions. Based on the current research, there are 18 ingredients used in Persian ink making which are identified to be colourants.

*3.1.1 Lamp Black—*دوده

Lamp black or *Dude* as it is known in Persian (Steingass 2004, p. 542) is basically soot or carbon black created by collecting soot from burning oil lamps. A thick wick made of cotton is inserted into a lamp filled with oil, placed under a hollow earthen container with a hole in the top. A plate is placed above the hole to capture the soot. Lamp black is then removed with the aid of feathers, wrapped in paper, secured in dough, and baked in a furnace in order to remove the residue oil (Heravī 16th C). This refers to the calcination (heating) process where some of the residual oil and impurities turn to carbon while the rest are removed in the form of gas or vapour (Mitchell and Hepworth 1904, p. 154), resulting in pure carbon, the main constituent of carbon ink. Lamp black is a very stable pigment unaffected by light, acids, and alkalis (Fig. 3.1).

*3.1.2 Gall Nuts (Tannic Acid Extract)—*مازو

Gallnuts, also known as Turkish galls or Aleppo gals or *Māzu* in Persian (Steingass 2004, p. 1140), are sphere like growths that develop on the branches, shoots and leaves of trees such as the oak tree, in response to parasitic insects like the female gall wasp puncturing an area and inserting its egg/s (Kolar and Strlič 2004, p. 149). Gallnuts contain a high amount of a specific tannin known as tannic acid ($C_{76}H_{52}O_{46}$) a type of polyphenol (Mitchell and Hepworth 1904, p. 45). It is a key ingredient in the manufacture of iron-gall ink, Persian ink and

Fig. 3.1 Lamp black—Image by Author

Fig. 3.2 Gall nuts—Image by Author

Peacock ink (Koroush 2011). On its own, the extract is not considered a colourant. With fermentation or hydrolysis, the extract releases glucose and gallic acid ($C_6H_2(OH)_3COOH$). However, when it is combined with different vitriols), a chemical reaction creates a water-soluble brown to purple iron-gall ink (iron gallate), which darkens with oxidation of the iron ions from ferrous (Fe^{2+}) to ferric (Fe^{3+}) to form a water insoluble complex. Therefore, the ink must be stored in a well-stoppered bottle to limit exposure to atmospheric oxygen (Fig. 3.2).

*3.1.3 Green Vitriol—*زاج سبز

Green vitriol is also known as iron vitriol, copperas, melanterite or Zāj-e sabz in Persian (Steingass 2004, p. 605). Chemically it is a hydrated iron (II) sulphate with the chemical formula $FeSO_4 \cdot 7H_2O$ (De Michele 1976, p. 59). In its natural form this mineral can be found in copper mines. When this metal sulphate is combined with the extract of gallnuts, a water-soluble iron-gall ink is created which upon oxidisation, creates a non-soluble blue-black complex (Krekel 1999, p. 54). It is also present in Persian and Peacock inks (Khoroush 2011) (Fig. 3.3).

In addition to acting as a colorant, green vitriol is also used as a mordant (Cardon 2007, p. 41). Other names for green vitriol mentioned in the treatises are Cyprus vitriol (Heravī 16th C), Turkish vitriol (Rostamdāri, 16th C), Khotan vitriol (Ḥalīyat al-kuttāb 16th C) and Kermani vitriol (Resāle dar Bayān-e Tariqe-ye Sāḵtan-e Morakkab-e Alvān va Kağaḏhā-ye Alvān, 16th C). These names reflect

Fig. 3.3 Natural green vitriol—Image by Author

Fig. 3.4 Iron sulphate (synthetic green vitriol)—Image by Author

the place and area where the vitriol comes from such as Cyprus or Roman vitriol (Krekel 1999, p. 55) which were confirmed by (Jokar 2015).

Occasionally the term white vitriol (*Zāj-e Sepid*) is mentioned, which is often mistaken for alum (hydrated potassium aluminium sulphate) by some contemporary calligraphers. However, alum does not react with tannic acid, and it is generally used as a mordant. Furthermore, the official Persian term for alum is *Zame* (ibid.). In such cases, the term white vitriol is identified to be a lesser hydrated form of green vitriol ($Fe(II)SO_4$). It should not be mistaken for zinc vitriol or zinc sulphate ($ZnSO_{4.}7H_2O$), which is also white in colour (De Michele 1976, p. 59) (Fig. 3.4).

3.1.4 Blue Vitriol—زاج کبود/آبی

Blue vitriol is also known as Zāj-e kabud/ābi in Persian (Steingass 2004, p. 605) and copper vitriol or *chalcanthite*. Chemically it is hydrated copper (II) sulphate with the chemical formula $CuSO_4{\cdot}5H_2O$ (De Michele 1976, p. 59). It is found in copper mines and reacts with tannins to create iron-gall ink. It is used in Persian and Peacock inks (Khoroush 2015) as well, and much like green vitriol, copper sulphate is also used as a mordent (Cardon 2007, p. 47) (Fig. 3.5).

Fig. 3.5 Copper Sulphate (synthetic blue vitriol)—Image by Author

3.1.5 Red Vitriol—زاج سرخ

In some treatises such as Ḥaliyat al-Kottāb (16th C), the use of red vitriol or *Zāj-e sorkh* in Persian (Jokar 2015) is mentioned instead of green vitriol. According to Jokar (2015) in a personal correspondence, red vitriol is potassium ferricyanide also known as red prussiate of potash or Prussian red with the chemical formula $K_3[Fe(CN)_6]$ (Fig. 3.6).

3.1.6 Yellow Vitriol—زاج زرد

Yellow vitriol or *Zāj-e zard* in Persian (ibid.) has also been mentioned in the treatises Ḥaliyat al-Kottāb (16th C) as a primary ingredient in ink making. In a personal correspondence, Jokar (2015) identifies this mineral as potassium ferrocyanide also known as yellow prussiate of potash, with the chemical formula $K_4[Fe(CN)_6]\cdot 3H_2O$ (Fig. 3.7).

3.1.7 Marcasite—مرقشیشا

Known as *Marqešišā* in Persian (Steingass 2004, p. 1218), Marcasite was the term used to describe bismuth (Bi), a metal similar to lead (Pb), also referred to as

Fig. 3.6 Potassium ferricyanide (synthetic red vitriol)—Image by Author

Fig. 3.7 Potassium ferricyanide (synthetic yellow vitriol)—Image by Author

tin-glass (Ure 1853). According to (Eastaugh 2008) there has been a lot of confusion over the meaning of marcasite in older literature. However, today Marcasite refers to a mineral with the chemical formula FeS (iron II sulfide). It is sometimes used instead of vitriol and added to the extract of gallnuts (Ḥalīyat al-kottāb 16th

CE). Another form of iron sulphide is pyrite (FeS_2) (Roberts, Rapp and Weber 1974, p. 381), also known as fool's gold due to its brass-yellow colour and metallic lustre (De Michele 1976).

3.1.8 Copper—مس

Copper (Cu) with the Latin name *Cuprum* (Syvum 2016), is known in Persian as *Mes* (Steingass 2004, p. 1224). It was used as a secondary colourant and possibly as a mordant in the manufacture of Peacock ink (Jokar 2015).

3.1.9 Verdigris—زنگار

Verdigris, is known as *Zangār* in Persian (Steingass 2004, p. 626). It is mostly referred to as copper acetate ($Cu(CH_3COO)_2.H_2O$) (Feller et al. 1993, p. 131), and is created when copper reacts with a source of acetic acid such as vinegar to form a patina, creating a beautiful green coloured pigment (Barkeshli and Ataie 2002, p. 155). This term is also used to describe the blue or green corrosion of copper. Like iron-gall ink, verdigris is known to degrade paper over time; however, it is suspected the addition of saffron prevents the destructive nature of Verdigris (Fig. 3.8).

Fig. 3.8 Verdigris—Image by Author

3.1.10 *Lapis Lazuli*—لاجورد

Lapis lazuli or *Lājvard* in Persian (Steingass 2004) is a natural occurring blue coloured stone composed of a mixture of minerals like calcite, pyrite and lazurite (Eastaugh 2008, p. 223). Lazurite ($(Na, Ca)_8(Al, Si)_{12}O_{24}(S, SO_4)$) (Roberts et al. 1974, p. 348) is found in large deposits in Badakhshan, Afghanistan (De Michele 1976, p. 114). This mineral is turned into powder by braying with a mortar and pestle and is used as a blue pigment and as a secondary colorant in Peacock ink (Khoroush 2011). The chemical composition of Lapis Lazuli is $(Na, Ca)_8(A1Si0_4)_6(S0_4,S,Cl)_2$ (Feller et al. 1993, p. 37) (Fig. 3.9).

3.1.11 *Green Walnut Husks*—پوست گردکان سبز

Walnuts are described as dry and warm natured and have been used as a food supplement and medicinally. Husk extract was traditionally used as an anti-inflammatory to treat inflamed throat and/or gums. Walnut husks beaten and boiled over a long period of time create a long-lasting dark colour and thus was used in ink making as a colourant (Khoroush 2015). It has also been used as a dye for colouring hair. Wool dying using fermented walnut husks was first documented in Persia (Cardon 2007, p. 78). Extraction involves leaving husks to ferment in water for two years before use, to maximise the dark brown colour. If fermented for only a few days, a

Fig. 3.9 Lapis lazuli pigment—Image by Author

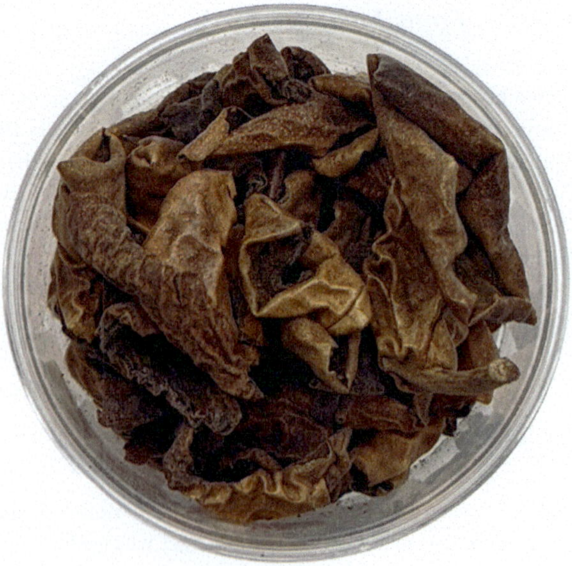

Fig. 3.10 Walnut husks—Image by Author

greyish green dye is obtained. To obtain a light brown (camel colour), in addition to dyeing with walnut husks, fibres were also submerged in an additional madder dye bath (ibid., p. 77) (Fig. 3.10).

3.1.12 Henna—حنا

Lawsonia inermis, also known as henna or *Ḥanā* in Persian (Steingass 2004) is a plant native to western India used to dye wool and silk fibres orange-russet colours. In regions of Iran such as Kerman, Bam and Baluchistan, henna is used with indigo to provide a fast-black colour (Cardon 2007, p. 85). In traditional Persian medicine, henna is known to have a cold and dry quality. Traditionally henna was used to heal scars and mouth ulcers, promote hair growth and dye hair. Drinking henna tea is believed to improve problems with jaundice, kidney and gall bladder stones and the spleen (Mir-Heidar 2012). Henna is considered to have both antibacterial and fungicidal qualities when mixed with water in 1 (henna) to 10 (water) proportions (Barkeshli et al. 2008). It has also been used since antiquity as a dye, a colourant in inks and possibly as a preservative (Fig. 3.11).

3.1.13 Indigo—نیل

Indigofera tincotoria or Indigo also known as *Nil* in Persian (Steingass 2004, p. 1444) is a distinctive blue dye. It is obtained from a plant that is native to

Fig. 3.11 Henna—Image by Author

many countries, especially on the Indian sub-continent, hence its name (Feller et al. 1997, p. 81). It also grows in northern Iran, especially in the province of Mazandaran and the foothills of the Alborz mountains. To extract the indigotin dye from the plant, its leaves and branches are submerged in water for 24 h. When oxidised by air, the dye becomes blue in colour (Mir-Heidar 2012). According to Jokar (2015) an Alkali is needed to extract the blue colourant.

Indigo is considered a dry and warm natured plant. In traditional Persian medicine, it is used as a blood coagulant. When mixed with other herbs, it is used to alleviate chest and kidney pains (Mir-Heidar 2012). It is used as a secondary colourant in inks (Khoroush 2011) (Fig. 3.12).

3.1.14 Woad—وسمه

Isatis tinctoria (woad), known as *Vasme* in Persian (Steingass 2004, p. 1468) and often called European indigo, was used as a blue dye for generations. It grows in the northern parts of Iran. According to traditional Persian medicine, woad is of a warm and dry nature and, much like henna, is used to promote hair growth and to treat common colds (Mir-Heidar 2012). Depending on the quality, fermented woad can achieve green, blue and black dyes (Storey 1978, p. 61). It is used as a secondary colourant in inks (Khoroush 2011) (Fig. 3.13).

Fig. 3.12 Indigo cake—Image by Author

Fig. 3.13 Woad—Image by Author

3.1.15 Myrtle—مورد

Myrtus communis or common myrtle, known as *Murd* in Persian (Steingass 2004, p. 1343) grows in temperate to warm weather and is native to Asia and southern Europe. In Iran, it grows in provinces such as Lorestan, Kerman, Bandar Abbas,

Baluchestan and Isfahan (Mir-Heydar 2012). Practitioners of traditional Persian medicine consider myrtle to be of a temperate quality and use it as a painkiller to treat headaches, stomach aches, earaches, tooth aches as well as joint and bone aches. Like henna and woad, it is used to promote hair growth. In a powdered form it is used as a deodorant. When applied to the skin, it is used to treat scars and ulcers (ibid.). It is used as a secondary green colourant in making Peacock ink (Khoroush 2011) (Fig. 3.14).

3.1.16 Saffron—زعفران

Crocus sativus is the scientific name for the species from which the expensive spice saffron is sourced, also known as *Za'firān* in Persian (Steingass 2004, p. 617). It is the stigma of the saffron flower, which is used in cooking and to extract the dye. On average, it takes 10,000 flowers to make 680 g of saffron. Practitioners of traditional Persian medicine consider saffron to be warm and dry natured. It is used to promote pleasure, cleanse kidneys and the bladder and to strengthen lungs. It is used as a secondary colourant in making Peacock ink (Khoroush 2011). Saffron is also a good buffering agent. When mixed with verdigris in Persian miniature paintings, it appears to have reduced the destructive degradation properties of Verdigris (Barkeshli and Ataie 2002, p. 162). Saffron may have been used in ink making for the same reason (Fig. 3.15).

Fig. 3.14 Myrtle—Image by Author

Fig. 3.15 Saffron—Image by Author

Fig. 3.16 Madder—Image by Author

3.1.17 Madder—روناس

Madder or Rubia Tinctorium or *Runās* in Persian (Steingass 2004, p. 597) is a plant which is native to the Mediterranean, Asia, northern Africa and the Caucasus. Within Iran, it grows in various areas such as Arak, Rei, Damavand, Tabriz and

Baluchestan. Root of Madder, which is generally used, is of dry and warm nature according to practitioners of Persian medicine and is generally used as a muscle relaxant and an anti-inflammatory. Mixed with honey, it is used to treat Jaundice and aids in digestion. Alizarin, the famous crimson red dye is extracted from the root of madder. In ink making, it is used as a secondary colorant, or it could be used as a primary colourant for making red ink (Khoroush 2011) (Fig. 3.16).

3.1.18 Aloe Vera—صبر

Aloe Vera, named *Ṣabr* in Persian (Steingass 2004) is normally used for its medicinal properties and as a preservative in inks. However, it is also used as a pigment when the juice is extracted and evaporated from the leaves, resulting in a yellow–brown coloured pigment (Eastaugh 2008, p. 11) (Fig. 3.17).

3.1.19 Cinnabar/Vermillion—شنجرف

Cinnabar is a naturally occurring red-scarlet mineral mercury (II) sulphide (HgS) also known as *Šanjarf* in Persian (Eastaugh 2008; Steingass 2004). It has been used as a pigment and as an additional colourant in making Peacock ink (Khoroush 2011). The synthetic form known as vermillion is used in European inks introduced by Ḥosseini (19th C) (Figs. 3.18 and 3.19).

Fig. 3.17 Dried aloe vera—Image by author

Fig. 3.18 Cinnabar—Image by Author

Fig. 3.19 Vermillion (synthetic cinnabar)—Image by Author

3.1.20 Gold—طلا

Referred to as *Ṭalā* or *Zar* (Steingass 2004, pp. 612, 816) in Persian, gold has the chemical symbol Au and Latin name *Aurum* (Syvum 2016). It has been used in powdered form (from leafgold) as a pigment mainly in watercolours (Eastaugh 2008, p. 177). In Peacock ink, it has been used as a secondary colourant and as a glossing (shining) agent (Khoroush 2011).

3.1.21 Silver—نقره

In Persian, silver is referred to as *Noqre* (Steingass 2004). It has the chemical symbol Ag derived from the Latin *Argentum* (Syvum 2016). It has been used in the form of silver leaf in illuminated manuscripts and in powdered form. At times, it is mixed with other pigments such as saffron to give a golden appearance (Eastaugh 2008, p. 348). With Peacock ink, it is used as a secondary colourant as well as a glossing (shining) agent (Khoroush 2011).

3.1.22 Tin—قلع

With the chemical symbol of Sn and Latin name *Stannum* (Syvum 2016) tin and tin compounds have been used as pigments. In powdered form it was used to imitate gold (Eastaugh 2008, p. 368). Relating to Persian inks, tin had been used in conjunction with mercury to create a form of metallic ink. In Persian, tin is referred to as *Qal'* (Steingass 2004).

3.1.23 Mercury—سیماب/جیوه

With the chemical symbol Hg and Latin name *Hydrargyrum* (Syvum 2016), compounds of mercury such as mercury iodide, mercury chromate, mercury sulfates and mercury sulphides had been used as pigments throughout history (Eastaugh 2008, pp. 264–265). Also known as quicksilver, mercury had been used in conjunction with tin to create a form of metallic ink in Persian ink making. In the Persian language, mercury is referred to as *Jiveh* and quicksilver is referred to as *Simāb* (Steingass 2004).

3.1.24 *Brass*—برنج

Brass is a golden coloured metal alloy made of copper and zinc. In powdered form, it was used as a form of metallic pigment on Japanese artefacts (Eastaugh 2008, p. 69). In Persian inks, it was used as an additional colourant in Peacock ink (Khoroush 2011). In Persian, brass is referred to as *Berenj*, a word also used to describe rice (Steingass 2004, p. 179).

3.1.25 *Red Poppy*—شقایق نعمان

Known as *Shaqāyeq-e Na 'mān* in Persian, this flower is from the genus Papaveraceae and naturally grows in areas high in lime. This plant is local to Asia, Europe, and most temperate climates above the equator. It is of dry and cold temperament according to traditional Persian medicine, and the petals and root of the plant is used to treat jaundice and alleviate throat and chest ache. In Persian ink making it is used as a red colourant.

3.1.26 *Red Mallow*—خیری سرخ

Known as *K̲eiri-e Sork̲* in Persian (Steingass 2004, p.492), or *K̲eiri-e Barri* (Mir-Heidar 2012), is a plant from the genus Cruciferae with the scientific name Mathiola Incana. In traditional Persian medicine, it is considered of dry and warm temperate and has a variety of medicinal uses: Eating the stem can cause miscarriage, and can help extract dead fetus, application of it can help in hair growth and acts as an anti-inflammatory when applied on muscles, and the leaves of it is used to alleviate tooth ache. In traditional Persian ink making, it is used as a red colourant.

3.1.27 *Sumac*—سماق

Known as *Somāq* in Persian and with the scientific name Rhus coriaria L, this plant is from the genus Anacardiaceae and is native to Iran. This plant is high in tannin and contains citric acid. According to traditional Persian medicine, sumac has a cold and dry nature. This plant is known to be astringent, good for the stomach, coagulant of blood, anti-diuretic. Due to its sour taste, it is used as a spice in Middle Eastern food, and it is often eaten to balance out oily food. In traditional Persian medicine it is used as a secondary red colourant, and due to high level of tannin, used as a replacement or addition to gall nuts (Fig. 3.20).

Fig. 3.20 Sumac—Image by Author

*3.1.28 Brazilwood—*بقم

Known as Baqam in Persian, Brazilwood is from the genus Caesalpinia. Brazilin is the main colourant that is extracted from this tree. Besides dye making, brazilwood is used as a red colourant in Persian ink making. In Persian treatises, there are multiple versions of Baqam. The term *Baqam* alone means Brazilwood while red baqam (*Baqam-e Sorḵ*) is sapanwood, and purple baqam (*Baqame-e Banafsh*) is logwood (Steingass 2004). Hence, Brazilwood, sapanwood and logwood might have been used interchangeably.

*3.1.29 Logwood—*بقم بنفش

Referred to as *Baqam-e Banafsh* in Persian (Steingass 2004, p.194), this tree is from the genus Leguminopsae and secondary family of Caesalpinaceae, with the scientific name Haematoxylon campechianum L. This plant is native to Mexico, and red dye is extracted from the wood known as Hematin. Moreover, it is a plant high in tannin as well as rosin. Traditional Persian medicine considers logwood as very warm and dry. The powdered wood is used to coagulate blood, and is applied to old and new scars, and washing face along with water would freshen the skin but should not be consumed. Logwood is used as a red colourant in both Persian, and European ink and colour making.

*3.1.30 Sappanwood—*بقم سرخ

Along with brazilwood and logwood, sapanwood known as *Baqam*-e *Sork̲* (red baqam) with the scientific name Biancaea sappan was used extensively in colour making, and as a red colourant in ink making.

*3.1.31 Indian Catechu—*کات هندی

With the scientific name Acacia catechu (L.) Willd and referred to as *Kāt-e Hendi* in Persian, this plant is from the genus Mimosaceae and is native to India, Ceylon, Burma, and west Africa. The wood and bark of catechu is high in tannin. The sap is a coagulant and is used to heal scars and treat diarrhoea. It is also used to treat chest pains and high fevers. The only recipe found using catechu to make red ink, is from the European recipes translated to Persian by Ḥosseini (19th C).

*3.1.32 Cochineal—*کوچنیل

Transliterated as کوچنیل in Persian and seen only once in a European ink recipe translated to Persian by Ḥosseini (19th C), cochineal is a red dye which is primarily extracted from species of scale insects known as Coccoidea. Some species are native to central and south America, while others are native to eastern Europe (Eastaugh 2008, p. 124) (Fig. 3.21).

*3.1.33 Turmeric—*زردچوبه

Turmeric is from the family of Zingiberaceae with the chemical formula Curcuma domestica val. This plant is known in Persian as *Zard-čube* (yellow wood) and is native to the west part of the Indian subcontinent, and south Asia. Practitioners of traditional Persian medicine consider turmeric to have warm and dry nature, and they use it as an anti-inflammatory, to treat toothache, and treating scars. It is used extensively in cooking, as a food dye, and in Persian ink making, used as a yellow colourant (Fig. 3.22).

*3.1.34 Minium/Red Lead—*سرنج

Known as *Soranj* in Persian, lead oxide with the chemical formula PbO is a pigment that has three colours: yellow, orange, and red (Eastaugh 2008, p. 234). The orange-red pigment seen in many manuscript paintings is also known as minium,

Fig. 3.21 Cochineal beetles—Image by Author

Fig. 3.22 Turmeric—Image by Author

which the term miniature painting comes from. As in a painting containing the pigment minium (red lead). Besides being used as a pigment, red lead is used to create red inks, and used as a secondary ingredient in Peacock ink, in traditional Persian ink making.

3.1.35 Antimony Collyrium—سرمه عملی

Known as *Sorme-ye 'Amali* in Persian, antimony or stibnite with the chemical formula Sb_2S_3, is a grey-black crystals with a metallic lustre that occurs naturally in metal deposits in places such as Turkey, Germany, England, Greece, Mexico, and Canada (Eastaugh 2008, p. 22). The crystals were traditionally used as an ailment for eye. A single recipe in *Resāle-ye Buqalamun* (Bahmani 15th C) uses this as a colourant in constructing black and white ink.

3.1.36 Orpiment—زرنیخ

Orpiment is a yellow to yellow-green colour pigment with the chemical formula As_2S_3 (Eastaugh 2008). Due to it containing arsenic, it is poisonous. It is known as *Zarnik̲* in Persian, and besides being used in paintings, it is used in the construction of yellow and red Persian inks and liqs (Fig. 3.23).

3.1.37 Asphaltum—اسفالتم

Asphaltum or asphalt with the Persian transliteration اسفالتم occurs naturally in sedimentary rocks, and it can be a by-product of the coal and petroleum industries and its main component is bitumen (Eastaugh 2008, p. 32). Only a single European recipe translated to Persian by Hosseini mentions this ingredient.

Fig. 3.23 Orpiment (www.masterpigments.com)

3.1.38 *Pumpkin Rind*—پوست کدو

Only one recipe from the treatise *Bayān al- Ṣanā ʻāt* by Teflisi (12th C) uses pumpkin rind (*Pust-e Kadu*) as an ingredient, and that is to burn it and turn it into charcoal, to then use as a main black colourant for ink.

3.1.39 *Lac*—لاک

Known as *Lāk* in Persian, lac is a red dyestuff extracted mainly from scale insects known as Coccoidea native to India, China, Cambodia, Muluccas, Sumatra and Thailand (Eastaugh 2008, p. 220). In Persian ink making, there was only one recipe found in *Resāle-ye Buqalamun* by Bahmani (15th C) which uses lac to make a pistachio-coloured ink.

3.1.40 *Chinese Blue/Prussian Blue*— چاینیز بلو / پرشن بلو

Chinese blue or Prussian blue with the chemical formula $[Fe(II)(CN)_6]^4$—is a synthesized blue colour that was constructed in the early eighteenth century (Between 1704 and 1707) and used as a pigment (Eastaugh 2008, p. 314–315). Amongst the recipes studied, only European ink recipes translated to Persian by Ḥosseini (19th C) mentions the usage of this pigment as a colourant (Fig. 3.24).

3.1.41 *Talc*—طلق

With the chemical formula $Mg_6Si_8O_{20}(OH)_4$ (Eastaugh 2008) this magnesium silicate hydroxoide is a greasy soft mineral which has been used extensively as an ingredient in red, gold, and silver liqs in the treatise *Resāle-ye Buqalamun* (Bahmani 15th C). In Persian, it is known as *Ṭalq* (Fig. 3.25).

3.1.42 *Lead Ceruse*—اسفیداج ارزیز

Lead ceruse or cerussite, a white lead carbonate with the formula $PbCO_3$ occurs during the oxidation of lead ore deposits while the synthetic form is commonly known as lead white (Eastaugh 2008, p. 97). In Persian, it is known as *Esfidāj-e Arziz*, and is used as a secondary colourant making yellow inks and yellow liqs.

Fig. 3.24 Prussian blue—Image by Author

Fig. 3.25 Talc powder—Image by Author

3.1.43 *Myrobalan*—هلیله

Known as *Halile* in Persian, Myrobalan is the fruit of a tree from the genus Combretaceae which grows in India, Ceylon, Burma, and the bottom of the Himalayas. Myrobalan contains tannin, is used as a blood coagulant and an anti-inflammatory for mouth ulcers and promotes gum health (Mir-Heidar 2012). In 1 recipe by Bahmani (15th C), it is used as a colourant to make green ink).

3.1.44 *Pigeon's Blood*—خون کبوتر

In a single recipe by Teflisi (12th C), pigeons' blood known as *K̲un-e Kabutar* in Persian is used to make sympathetic (invisible) ink with the addition of urine.

3.1.45 *Bitter Orange*—نارنج

Known as *Nārenj* in Persian, the juice of bitter orange was used as the main ingredient in making sympathetic ink in 1 recipe by Teflisi (12th C). Bitter orange with the scientific name Citrus bigaradia L (Mir-Heidar 2012) has a sour juice which is cold and dry in its temperament according to traditional Persian medicine and is used to treat coughs and phlegm.

3.1.46 *Urine*—آب تاختن

Only 1 recipe was found by Teflisi (12th C) that uses urine known as *Āb-e Tāk̲tan* to make a sympathetic ink.

3.1.47 *Fish Oil*—روغن ماهی

Only 1 recipe was found by Teflisi (12th C) that uses fish oil known as *Rog̃an-e Māhi* to make a sympathetic ink.

*3.1.48 Colocynth—*حنظل

Known as *Ḥanẓal* in Persian, colocynth or bitter cucumber with the scientific name Citrullus colocynthis (L.) Schrad is mentioned once in a recipe on sympathetic ink by Teflisi (12th C). According to traditional Persian medicine, Colocynth is extremely warm and dry in nature, and is used as a laxative.

Most of the above-mentioned colourants were used as pigments and dyes in illustrations and coloured inks. Some like saffron, henna, verdigris, indigo and lapis lazuli were also traditionally used to dye paper (Barkeshli 2016).

3.2 Binding Media

A binding medium is often used in ink making to disperse the colourant evenly within the vehicle (generally water) and prevents the colourant from precipitating at the bottom of the ink solution. In addition, the binder helps to attach the colourant onto the support (paper) to which it is applied. The treatises researched showed five ingredients that were used in Persian ink making as binding mediums.

*3.2.1 Gum Arabic—*صمغ عربی

Gum Arabic or *Ṣamq-e ʻArabi* as it is known in Persian (Steingass 2004) is the sap or polysaccharide which exudes from branches of the Acacia Senegal tree when the bark surface has been pierced (Mir-Heidar 2012) by insect bites, branches grinding against each other or when it is scratched deliberately.

Depending on the type of tree, time of the year and method of extraction, the gum can have shades from white to brown. Four types of gum Arabic (coloured white, yellow, red and brown) are mentioned by Jokar (2015). The darker the gum, the lower the quality. Gum Arabic dissolves in water and contains the sugars arabinose ($C_5H_{10}O_5$) and galactose ($C_6H_{12}O_6$). In traditional Persian medicine, gum Arabic is known to be temperate and dry and is used to treat throat and chest pains, coughs, and diarrhoea. It is also used as a sizing and adhesive in the textile and paper manufacturing industries and as a binding medium in inks (Khoroush 2011; Khoroush 2015; Jokar 2015) (Fig. 3.26).

*3.2.2 Liquidamber—*عنبر اشهب/عبیر اشهب

Liquidambar or *ʻAbir-e Ašhab/ʻAnbar-e Ašhab* in Persian (Steingass 2004, p. 834), also called sweet gum or red gum, is a polysaccharide extracted from the trees

Fig. 3.26 Gum Arabic pearls—Image by Author

of the Hamamelidaceae family. When used in traditional Persian medicine, it is described as having a very dry and warm quality. It is used as a relaxant for stiff muscles, and to treat coughs, tetanus, catarrhal, extreme shakes and tremors, as well as chest and throat aches (Mir-Heidar 2012). It is used as an additional binding medium or as an aromatic in inks (Khoroush 2015; Jokar 2015). It should be noted the Persian term *'Anbar'* is could also referred to Ambergris which is an excrement from marine animals such as the sperm whale (Steingass 2004, p 869) (Fig. 3.27).

3.2.3 Sandarac—سندروس

Referred to as *Sandarus in* Persian (ibid), sandarac is a yellow-white resin, somewhat transparent, aromatic and brittle. It is extracted from the sap of the sandarac tree, which is from the Genus *Tetraclinis articulata*, similar to cypress trees. Sandarac is categorized as a cupressaceae resin, which belongs to the diterpenoid resin family. It consists of polymerized communic acid (close to 70%) and sandaracopimaric acid ($C_{20}H_{30}O_2$). In the west, it is used as a varnish. In Persian ink making, it is used as a secondary binding medium (Khoroush 2011). Sandarac should not be confused with sandaraca or sandarach, which are orpiment and realgar respectively (Eastaugh 2008, p. 338) (Fig. 3.28).

Fig. 3.27 Liquidamber—Image by Author

Fig. 3.28 Sandarac—Image by Author

3.2.4 *Egg white*—سپیده خایه

Egg white known as *Sepide-ye Ḵaye* in classic Persian or *Sefide-ye Toḵm-e Morḡ* in modern Persian was occasionally used as a binding medium instead of gum Arabic, or as an additional minding media.

3.2.5 *Gum Ammoniac*— وشه / وشق

Known as Voše or Vošaq in Persian, this gum-resin was occasionally used instead of gum Arabic as a binding medium (Fig. 3.29).

3.3 Thickening Agents

Depending on the calligrapher, the ink created can at times be too fluid and subsequently, thickening agents are added to make the ink more viscous. Five ingredients have been identified as thickening agents in traditional Persian ink making.

3.3.1 *Camel Thorn*—برگ خار/ خار شتر

Alhagi Maurorum, also known as Camel Thorn or *Bargi ḵhār / ḵhār shutur* in Persian and as Alhagi in French, derived from the Arabic word Alhaji, literally meaning pilgrim. It is grown mostly in the southern parts of Iran such as Baluchistan, Bushehr, Kerman and Ahvaz. In traditional Persian medicine, Camel Thorn has a very warm and dry quality. The plant extract is used to treat kidney and gall bladder stones. Oil extracted from its leaves is good for rheumatism and the flower extract is used to treat haemorrhoids (ibid). It is used as a thickening agent in inks (Khoroush 2011) (Fig. 3.30).

3.3.2 *Dog Rose Extract*—عرق نسترن

Dog Rose or *'Araq-e nastaran* in Persian with the scientific name *rosa canina* grows on the outskirts of the Alborz and Zagrus mountains (Mir-Heidar 2012). In traditional Persian medicine dog rose is known for its dry and warm nature. It is used to help clot bleeding from scars, heal extreme diarrhoea and alleviate pains arising from kidney stones (ibid). It is used as both an aromatic (due to its pleasant aroma), and a thickening agent in inks (Khoroush 2011, 2015).

Fig. 3.29 Gum ammoniac (https://shop.stuartstevenson.co.uk/products/gum-ammoniac)

Fig. 3.30 Camel thorn—Image by Author

3.3.3 Epithymum—افتیمون

Cuscuta Epithymum or Dodder of thyme is known in Iran as *Aftimun* (Steingass 2004). It is a type of parasitic weed that grows in Iran around cities such as Tehran, Karaj, Shush, Hamedan and Isfahan. Epithymum is known to be of warm

Fig. 3.31 Epithymum—Image by Author

and dry nature in traditional Persian medicine. It is used to treat nerve related illnesses, dementia and psychosis, hypochondria, depression, and nightmares; and is called ‘Davā’ al-Junun’ meaning “the cure for madness”. It is used as a thickening agent in inks however, some master calligraphers advise against its use due to it compromising the smoothness of the ink (Heravi, 16th C-b) (Fig. 3.31).

*3.3.4 Coral—*کفک دریا

In powdered form, coral or Kafak-e daryā literally meaning ‘bottom of the sea’ in Persian, has been used as a thickening agent in inks (Heravi, 16th C-b). It is comprised primarily of calcium carbonate with additional trace elements. Many master calligraphers advised against using too much of materials like coral and epithymum as it reduces the smoothness of the written ink (ibid) (Fig. 3.32).

*3.3.5 Gum Arabic—*صمغ عربی

Due to its tacky and thick nature, gum Arabic and other binding mediums could also be used as thickening agents.

Fig. 3.32 Dried coral—Image by Author

3.4 Aromatics (Fragrances)

Due to its tacky and thick nature, gum Arabic and other binding mediums were also used as thickening agents. Aromatics were added when an ink had a less than desirable smell or the calligrapher wanted his ink to have a nice aroma, resulting in a better writing experience. In such cases, some ingredients were used for their aromatic (fragrant) qualities of which five ingredients have been identified.

3.4.1 Rose Water/Damask Rose Extract—گلاب

Damask rose or *Rosa Damescena* commonly referred to as Golāb in Persian (Steingass 2004, p. 1093), can be found in the Kashan, Fars, Meimand, Azerbaijan and Qazvin provinces of Iran. This particular type of red rose extract is appreciated for its strong and pleasant aroma. It is also used in the food and beverage industry. In ink making, it is used both as an aromatic, and to make the ink more fluid (Khoroush 2011, 2015).

3.4.2 Chicory Extract—عرق کاسنی

Chicory (*Cichorium Endivia*) or *'Araq-e kāsni* in Persian (Steingass 2004, p. 1004) is known to be of cold and wet quality. It has been used alone or in conjunction with other herbs in traditional Persian medicine to cure illnesses resulting from extreme heat such as headache, inflammation, and dehydration (Mir-Heidar 2012). Within Iran, chicory grows in many places such as the foothills of the Alborz Mountains, Qazvin, Rasht, Tehran, Orumieh, Lahijan, Fars, Baluchistan, Abadan and Khorasan (ibid). In ink making, it is used as an aromatic and possibly to make the ink more fluid (Khoroush 2011).

3.4.3 Musk—مشک

In many ink recipes, the term *Mošk* in Persian (Steingass 2004) or Musk is mentioned to be added in order to provide a nice aroma. Musk can have multiple meanings, such as the: glandular secretions of musk deer and similar animals; buds and flowers of aromatic trees such as the sweet acacia tree; musk seed or amber seed; or artificially made aromatics (Mir-Heidar 2012). In ink making it is used to give the ink a pleasant aroma (Khoroush 2011, 2015) (Fig. 3.33).

3.4.4 Liquidamber—عنبر اشهب/ عبیر اشهب

In addition to being used as a secondary binding medium, liquidambar has also been used as an aromatic in ink making due to its pleasant aroma (ibid).

Fig. 3.33 Plant-based musk—Image by Author

3.4.5 Dog Rose Extract—عرق نسترن

Besides being used as a thickening agent, dog rose was also used as an aromatic in ink making (ibid).

3.5 Preservatives

In order to protect ink from biological attack (bacterial, fungal, insect), the following three ingredients were identified to be used as preservatives in traditional Persian ink making.

3.5.1 Aloe Vera—صبر

In addition to its usage as a colourant, Aloe Vera was also used as a preservative. It is native to Africa and tropical countries and varieties can be found in the Baluchistan province of Iran. Extract or gel of Aloe Vera has been traditionally used to treat fever, scars, burns, pains related to menstruation or pregnancy, toothache, and stomach aches. Due to its alkaline properties, Aloe Vera is used as an antacid. It is also well known to have antibacterial and fungicidal properties (Mir-Heidar 2012). Its bitter taste has been used to dissuade toddlers from nursing and children from the habit of licking their lips (ibid). These same properties have been used in ink making to repel insects such as termites and to prevent mould growth (Jokar 2015).

3.5.2 Salt—نمک

Salt or sodium chloride also known as *Namak* in Persian (Steingass 2004, 1426), is well known for its antibacterial nature. Gargling salt water was used to reduce swelling and battle infections. In ink making, it is added as an antibacterial agent and to prevent fungal and mould growth although too much of it is not advised (Khoroush 2015).

3.5.3 Ox-Gall—زهره گاو

Due to the foul smell of ox-gall or *Zahr-e gāv* as it is called in Persian, this ingredient keeps insects away which is why it had been added to inks by master calligraphers (Khoroush 2015).

3.5.4 *Henna*—حنا

In addition to being used as a colourant, henna was also used as a preservative. Henna is known to have both antibacterial and fungicidal qualities if used in a 10% solution applied to paper (Barkeshli et al. 2008). It is probable that it was used for this reason in addition to its colourant qualities.

3.6 Glossing Agents

Calligraphers often valued the shine of an ink, which provides a more pleasant experience for the reader. Consequently, certain materials were introduced to add gloss to an otherwise matte writing media. There are four ingredients identified as glossing agents used in traditional Persian inks.

3.6.1 *Sugar*—شکر

Sugar or *Šekar* in Persian was added to inks in order to give black writing matter gloss (Khoroush 2015).

3.6.2 *Honey*—عسل

Honey, due to its high sugar content, much like sugar and sugar candy and rock sugar, was used to give gloss to the ink (ibid). In Persian honey is referred to as *'Asal.*

3.6.3 *Egyptian Sugar Candy*—نبات مصری

In many recipes (Heravi 16th CE; Sabzevāri 16th CE; *Majmu ' al-Ṣanāye '* 16th CE), Egyptian Sugar Candy or *Nabāt-e Meṣri* in Persian is mentioned, which is sugar that has been cooked and hardened into a form of candy.

3.6.4 *Gold*—طلا

Due to the reflective nature of gold, it was added to peacock ink to give it extra gloss and as an additional colourant (Khoroush 2011).

3.6.5 *Silver—*نقره

Similar to gold, silver was added to peacock ink due to its reflective nature and as an additional colourant (ibid).

3.7 Mordants

Mordants are typically inorganic oxides which aid the fixing of colourants to paper fibres (Cardon 2007, p. 20) as opposed to binders which loosely bind the colourant to the surface of the paper. Mordants increase the fastness of inks and can enhance the colour or shade of the dye applied (Bechtold and Mussak 2009, p. 30). In the fabric dyeing industry, mordants are either applied to the fabric prior to, during, or post dyeing (Storey 1978, p. 66). Two ingredients were identified to be used as mordant additives in Peacock ink recipes.

3.7.1 *Alum—*زمه

Hydrated potassium aluminium sulphate ($AlK(SO_4)_2 \cdot 12H_2O$) or Alum was advised by master calligrapher Solṭān ‘Ali Mašhadi (15th C) to be used instead of vitriol to create Persian ink. This mineral does not react with the extract of gallnuts and acts as a mordent to bind the ink more closely to the paper fibres (Cardon 2007, p. 21). In Persian the term for alum is either *Zame* or *Zāj-e Sefid*, meaning white vitriol. This should not be mistaken for zinc vitriol or zinc sulphate ($ZnSO_{4.}7H_2O$) described by De Michele (1976, p. 59), which is also white in colour (Fig. 3.34).

3.7.2 *Sal Ammoniac—*نوشادور

Ammonium chloride (NH_4Cl) or sal ammoniac is a mineral used as a mordant in conjunction with other materials such as tin (Ure 1853, p. 1258) and was possibly used in making Peacock ink for the same reason (Jokar 2015). In personal correspondence with Dr. Jalil Jokar (2015), it is mentioned that sal ammoniac is also used as an alkali to extract the blue dye from indigo leaves, and as a catalyst for chemical reactions in general. The Persian term for sal ammoniac is *Nushādur* (Steingass 2004, p. 1433) (Fig. 3.35).

Fig. 3.34 Alum powder—Image by Author

Fig. 3.35 Sal ammoniac—Image by Author

3.7.3 Green Vitriol—زاج سبز

Besides being used as a colourant when reacting with the extract of gallnuts, green vitriol was also used as a mordant (Cardon 2007, p. 41).

3.7.4 *Blue Vitriol*—زاج کبود/آبی

Besides being used as a colourant when reacted with the extract of gallnuts, blue vitriol was used as a mordant (ibid, p. 47).

3.7.5 *Marcasite*—مرقشیشا

Besides being used as a colourant when reacted with the extract of gallnuts, Marcasite could have been used as a mordant.

A study of the materials and techniques of Persian inks, and the technological and cultural considerations involved in making decisions about their use, provides information which substantially increases an understanding of the process of manufacture of Persian manuscripts, and assists in the scientific analysis and conservation treatment of significant cultural documents.

Chapter 4
Persian Ink Recipes

Ink recipes were gathered through researching treatises, translated from classic Persian to English and categorised. When comparing the different recipes, it was evident that some are very simple and give little information regarding measurements, (like carbon ink recipes) while others are extremely detailed (most peacock ink recipes). Regardless, there are challenges faced when translating the recipes and trying to equate the measurements of the constituents. Four terms are used to describe ink in the literature on Persian inks and ink making: *Morakkab, Medād, Siāhi* and *Johar.*

- *Morakkab* (مرکب) is of Arabic origin and means 'mixture', as ink is made from a mix of multiple ingredients.
- *Medād* (مداد) is derived from the Arabic term *Madad,* which means 'aid' or 'assistance', as it aids the pen in writing (Soltany et al. 2015, p. 9).
- *Siāhi* (سیاهی) is derived from the Persian term *Siāh* meaning 'black' since most o main writing matters are black (or dark) in colour.
- *Johar* (جوهر) is derived from Arabic and means 'gem'. It is commonly used in modern Iran to describe ink due to it being valued and there being so many varieties of different colours in modern times.

Some Persian words have general or multiple meanings. For example, *Zāg* or *Zāj*, which is generally translated to vitriol, can also translate to alum as seen in Porter (1985, p. 64), while the term *Zame* which is translated to alum by Minorsky (1959) could also be translated to vitriol (Steingass 2004). Another issue is that the word *Zame* is only mentioned twice in all the treatises researched but the word white vitriol (*Zāj-e Sefid*) is used more than a few times. It is a general understanding amongst contemporary Persian calligraphers today, that white vitriol and alum are synonymous (Koroush 2011) while Agricola (1950) defines white vitriol to be Zinc Sulphate ($ZnSO_4$). Another possibility is that white vitriol is a lesser hydrated form of green vitriol (Jokar 2015). In such cases the recipe should be studied and if there is any mention of chemical reaction, it is most likely green vitriol as alum does not react chemically with tannic acid extracted from gall nuts.

S. Zekrgoo, *Tradition and Science of Persian Ink Making*,
https://doi.org/10.1007/978-3-031-52071-6_4

Other coloured vitriols are occasionally mentioned such as red and yellow. According to Jokar (2015) the former (red vitriol) is potassium ferricyanide also known as red prussiate of potash or Prussian red with the chemical formula $K_3[Fe(CN)_6]$. The latter (yellow vitriol) is potassium ferrocyanide, also known as yellow prussiate of potash, with the chemical formula $K_4[Fe(CN)_6]\cdot 3H_2O$. Another possible interpretation for red vitriol could be the product from heating ferrous sulphate ($FeSO_4$) creating 'Copperas Red' or 'Burnt Roman Vitriol'(Eastaugh 2008), that is oxidised iron (III) sulphate ($Fe_2(SO_4)_3$) which has a red tone. Of the two possible explanations, the former is more probable as it occurs naturally.

Ṣabr means 'aloe'(Steingass 2004, p. 781). But the word *Ṣabr-e sarv,* in a Peacock ink recipe, translates to 'aloe of cypress' which in this instance could either mean gall-nut or the cones of a cypress tree.

One of the challenges with translating the recipes is Persian units of measurement and interpreting quantities into SI units (International system of units). These units may have changed over time and some of the units have multiple meanings. I had previously used measurements adopted from those used by herbal and jewellery stores at Bazaars in modern day Iran (Rasekhoon 2014; Bahesab 2015; Zekrgoo 2014; Zekrgoo et al. 2017). Fortunately, a recent study was done on the units of measurements used specifically in ink making, within Persian calligraphic treatises by Azadi-Buyaghchi, Afsharpur and Haddadi (2017). The list of classical Persian units of measurements and their equivalent metrics based on said research is given in the following table which is used throughout this book (Table 4.1).

Table 4.1 Classic persian units of measurement according to Azadi-Buyaghchi, Afsharpour and Haddadi (2017)

Name in persian	Transliteration	Equivalent metric unit
دانگ	Dāng	2.5 g or a 6th of a *Deram*
مثقال	Meṯqāl	4.64 g
درم	Deram	15 g
رطل	Raṭl	389.76 g, or 84 *Meṯqāl* or12 *Uqiah*
سیر	Sir	75 g, or 5 *Deram,* or 16 *Meṯqāl*
چارک	Čārak	750 g or 10 *Sir*
اوقیه	Uqieh	32.48 g or 7 *Meṯqāl*
نخود	Noḵod	4.6 g
توله/ تولچه	Tule/Tulče[1]	16.2 g or 3.5 *Meṯqāl*
من	Mann[2]	$\approx$ 3 kg Tabriz = 2.96 kg Mashhad = 3.05 kg Kerman = 3.1 kg Shiraz = 3.3 kg Qazvin = 1/6th of a kg

[1] This unit of measurement originates in India, and like Mann, had different measurements in different regions. During the Qajar period of Iran, it was equivalent to 3.5 *Meṯqāl.*
[2] *Mann* has different weights according to different region. Most were close to 3 kg, so this measurement was used in personal ink reconstructions.

Each recipe is given a prefix letter 'R' and a number to represent the number of recipes available for each category of ink i.e., R1, R2, R3 etc. A recipe from each ink category was selected for reconstruction at the Grimwade Centre laboratory, The University if Melbourne, followed by subsequent analysis. This was based on each recipe providing sufficient information (regarding ingredients, quantities, instructions, guidelines etc.) and the ingredients involved being available for purchase. Other recipes were not selected due to a lack of information and due to unacceptable Occupational Health and Safety (OH&S) risks associated with the ingredients or preparation method.

The aim was to re-construct these historic inks based on the original recipes provided. However it should be noted that the re-constructed inks will never be 100% historically accurate for the following reasons: availability of source material; differences in where the material is acquired from; accuracy of the recipes; whether the recipes are in their intended language or if information was lost in translation; general terminologies may have multiple meanings; differences between tools used during original construction in the past versus modern tools used to re-construct inks; and other factors.

t's noted that the reconstruction of numerous ink recipes has been undertaken, yet the detailed coverage of these reconstructions will be reserved for the second volume, currently in progress. The forthcoming volume will delve into the intricacies of reconstructing the inks based on the elucidated recipes in this book.

4.1 Black Inks

Black ink is a general term used to describe dark inks which are used as the primary writing matter as opposed to coloured inks which were often used for title headings, translations under the main text (e.g., Qur'an), and highlighting or emphasising certain words. As mentioned before, one of the terms to describe ink in Persian treatises is *Siāhi* (سیاهی) meaning black matter.

4.1.1 Carbon Ink

Carbon ink, the oldest and most common ink used in history is made from a suspension of carbon particles (made of soot) and a binder (gums or glue). The source of carbon in Persian inks were almost always lamp black, and the binder was gum Arabic. Due to its simplistic nature, most calligraphers knew how to create this ink. Hence, few recipes explain how to make carbon ink. However, other ink recipes used by Persian master calligraphers incorporated carbon ink as one of the main constituents. Therefore, methods for creating carbon ink are provided in the initial stages of these recipes.

The first step in creating carbon ink involves collecting lamp black by inserting a thick wick made of cotton, into a lamp filled with oil and placing the lamp under a hollow earthen container with a hole at the top. A cold plate is placed above the hole to capture the soot. Lamp black is then removed from the plate with the aid of

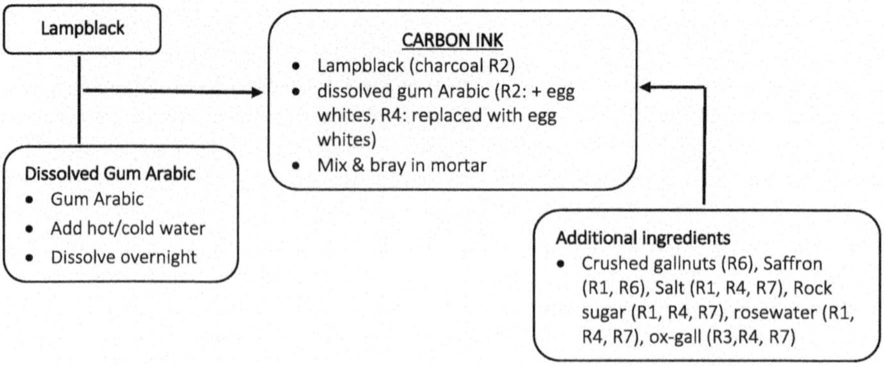

Fig. 4.1 Flowchart guide for constructing carbon ink

feathers, wrapped in paper, secured in dough, and baked in a furnace so the residue oil can be burnt away through the calcination process.

Seven recipes were found on the creation of strictly carbon ink. Three are from the treatise *Bayān al- Ṣanā 'āt* (12th C), two are from the treatise *Majmu ' al-Ṣanāye ' (Collected Crafts* 16th C), one is from the treatise <u>*Resāle dar bayān-e Kat va Morakkab va Kaḡad va Sāktan-e Ranghā*</u> (15th C), and a final one from the treatise <u>*Golestān-e Honar*</u> (16th C) (Anonymous 15th C). For an overview of the constituents and measurements refer to Table 3.2. English translations of these recipes can be found below (Fig. 4.1; Table 4.2).

R1—Recipe from the treatise *Bayān al- Ṣanā 'āt* by Teflisi (12th C):

بستاند دوده چراغ نفت چندان که خواهد و اورا به هاون بساید چندانکه تابنده گردد، پس آب صمغ عربی صافی کرده بگیرد چندانکه نه ستبر باشد و نه تنک، و اندک اندک از وی به هاون بر مداد همی ریزد و نیکش همی ساید تا قوی گردد. پس لختی نبات با شکر بیامیزد و لختی نمک را به گلاب بگدازد و اندکی زعفران با وی بیامیزد و این جمله را با مداد دیگر باره بهم نیک بساید و در شیشه آبگینه کند و هنگام حاجت به کار برد.

Gather lamp black as much as desired and pound it in a mortar until lustrous, next acquire clean gum Arabic liquid to the point that it is neither too thick nor too thin and add to the mortar through drops and bray and mix well. Afterwards take some sugar, a piece of rock sugar, some salt, some saffron, and mix with rosewater, and pour this concoction into the mortar and bray well. Finally pour the ink in a glass container and used as desired.

R2—Recipe from the treatise *Bayān al- Ṣanā 'āt* by Teflisi (12th C):

بستاند زگال کدو و خردش بسر شد و با آمیخته وی را بیامیزد و بسپیده خایه بسر شد. پس بسایه بنهد تا خشک شود و با هر منی ده درم سنگ صمغ عربی بیامیزد و به آب مازو وئ را تر کند و در دویت ریزد و بکار برد.

Table 4.2 Carbon ink—ingredients and measurements

Carbon ink	Primary ingredients	Parts	Amount	Secondary ingredients	Parts	Amount
R1 (12th C)	Lamp black	–	–	Sugar	–	–
	Gum Arabic	–	–	Rock sugar	–	–
	Water	–	–	Salt	–	–
	–	–	–	Saffron	–	–
	–	–	–	Rosewater	–	–
R2 (12th C)	Pumpkin charcoal	–	3 kg	–	–	–
	Egg whites	–	–	–	–	–
	Gum Arabic	–	150 g	–	–	–
R3 (12th C)	Lamp black	–	3 kg	Ox-gall	1 piece	–
	Gallnuts	–	150 g	–	–	–
	Burnt paper	–	75 g	–	–	–
	Egg whites	–	–	–	–	–
R4 (15th C)	Lamp black	–	–	Salt	–	–
	Gum Arabic	–	–	Rock sugar	–	–
	Water	–	–	Rose water	–	–
	–			Ox-Gall	–	–
R5 (16th C)	Lamp Black	–	–	–	–	–
	Gum Arabic	–	–	–	–	–
	Water	–	–	–	–	–
R6 (16th C)	Lamp black	–	46.4 g	Gallnuts	A fourth	–
	Gum Arabic	A fourth	–	Saffron		2.5 g
	Water		–	–	–	–
		–			–	
R7 (16th C)	Lamp black	–	–	Salt	–	–
	Gum Arabic	–	–	Rock sugar	–	–
	Water	–	–	Rose water	–	–
	–			Ox-Gall	–	–

Gather pumpkin rind charcoal and beat it to fine powder and mix it with egg whites. Let it dry in shade; for every *Mann* (3 kg) of this add 10 *Deram* (150 g) gum Arabic, finally add gallnut extract, place in an inkwell, and use as desired.

R3—Recipe from the treatise *Bayān al- Ṣanā ʻāt* by Teflisi (12th C):

بستاند انقاس و بر هر منی ده درم سنگ مازوی سوده و پنج درم سنگ کاغذ سوخته و همه را به هاون نیک بساید و به
سپیده خایه سر شد و قرصهای خرد بکند و بکار برد. و اگر خواهد مگس بر دویت یا بر نبشته ننشیند لختی ز هره گاو اندر
دویت کند.

Acquire lamp black and for each *Mann* (3 kg), gather 10 *Deram* (150 g) powdered gallnuts and 5 *Deram* (75 g) burnt paper. Grind all these together in a mortar and add egg white and pound again; create small cakes and use when desired. If you want to prevent flies from gathering on the ink or the writing, add a piece of ox-gall in the inkwell.

R4—Recipe from the Treatise *Resāle dar bayān-e Ḵaṭ va Morakkab va Kāḡaḏ va Sāḵtan-e Ranghā* by Anonymous (15th C):

بستاند دوده چراغ، و در هاون کند و بساید که سخت تابنده شود. پس آب صمغ عربی صاف کرده بگیرد چندان که نه ستبر شود و نه تنک، و اندک اندک در هاون ریزد و نیک می ساید تا قوی گردد. و قدر نبات یا طبرزد و نمک به گلاب بگذارد و اندکی با وی بیامیزد. و این جمله را با مداد دیگر بساید نیک، و در شیشه کند و در وقت حاجت بکار برد. و اگر خواهد که مگس بر وی ننشیند اندکی زهره گاو در دوات اندازد.

Obtain soot and place it in a mortar and bray until radiant. Mix gum Arabic with water to the point that it's neither too thick nor too fluid and pour in small portions while braying well till it gets form. Add some salt and rock sugar to rose water, mix and add some to the ink and bray well. Pour the ink in a bottle and use as necessary. If you want to keep flies and insects away from the ink, add some ox gall.

R5—Recipe from the Treatise *Majmu' al-Ṣanāye'* by Anon (16th C):

بگیرند روغن کتان و در چراغدان کنند و فتیله سطبر کنند، و بگیرند خمره بزرگ، روی وی گشاده، چراغدان در آن خمره نهند چنانکه بر سر چراغدان باژ گونه نهاده باشند و کون خمره سوراخ بود، و خمره دیگر بر سر آن سوراخ گرداند و شیر در خمره بالایی بمالند و یک شبانروز بگذارند تا دوده از آن سوراخ از خمره بالایی بگیرد، پس آن به پر مرغ فرو گیرند و دوده که حاصل شده باشد در طاس برنجی کنند قطره قطره آب صمغ عربی در وی ریزند و به دست مالند، بنهند تا خشک شود چنانچه بترقد، پس فرود کند هر گاه که حاجت افتد در گوش ماهی به آب صمغ بسرشند و به کار برند هرجا که خواهند.

Purchase some linseed oil, place it in a lamp and make a thick wick. Use a big earthen container, turn it over and place it on top of the lamp; there must be a hole at the bottom of the container. Rub some milk at the bottom of another container, turn it over and place it above the first container and wait for one day and one night so that the lamp black is stuck to the top container. With the help of a feather, scrape off the soot and place it in a brass mortar. Add gum water to the soot in drops and rub and mix well with your hands until it takes the form of a paste and repeat. Afterwards let the paste dry and break into pieces. When needed, you may mix this dried ink with some gum Arabic water in a sea-shell and use as desired.

R6—Recipe from the Treatise *Majmu' al-Ṣanāye'* by Anonymous (16th C):

در صنعت مرکب مهوش: بگیرند مازو چهار یکی، و صمغ عربی چهار دوم،و دوده ده مثقال و زعفران دانگی، و به مانند پیش عمل کنند، نیکو آید و پسندیده.

On making *Mahvaš* (Beautiful) ink: Acquire a fourth of gall, a fourth of gum, 10 *Meṯqāls* (46.4 g) of lamp back and a *Dāng* (2.5 g) of saffron and follow the steps of the previous recipe (iron gall ink R2). It shall be a good and pleasant ink: Break the gallnuts into small pieces and pour water on top and set to boil until the liquid is evaporated to half its size. Then filter through cloth and let it cool down. Add saffron and lamp black and mix and bray the ink well while adding beaten and powdered gum Arabic to it slowly.

R7 Recipe from the treatise *Golestān-e Honar* by Qāżi Aḥmad Qomi (16th C):

بستاند دوده چراغ و در هاون کند و بساید که سخت سائیده شود پس آب صمغ عربی صاف کرده بگیرد چنانکه نه سطبر شود و نه تنک و اندک اندک در هاون ریزد و نیک بساید تا قوی گردد پس لختی نبات و طبرزد و نمک بگلاب بگذارد و اندکی باوی بیامیزد و این جمله را بامداد دیگر بساید نیک در شیشه کند و در وقت حاجت بکار برد اگر خواهد که مگس بروی ننشیند اندکی زهره گاو در دوات اندازد.

> Obtain soot and place it in a mortar and bray until radiant. Mix gum Arabic with water to the point that it's neither too thick nor too fluid and pour in small portions while braying well till it gets form. Add some salt and rock sugar to rose water, mix and add some to the ink and bray well. Pour the ink in a bottle and use as necessary. If you want to keep flies and insects away from the ink, add some ox gall.

It seems like Qāżi Aḥmad took this recipe from the treatise *Resāle dar Bayān-e Ḵaṭ va Morakkab va Kāġaḏ va Sāḵtan-e Ranghā* (15th C) since it is the exact same recipe as R4.

Carbon ink is a stable ink since no chemical reactions taking place between the constituents. The main concern with this ink is that it is possible to abrade and wash away the ink from the surface of the paper. This susceptibility is due to the ink sitting on top of the substrate. As the manuscript ages, the bond between ink and paper becomes stronger, making it harder to remove the ink manually.

4.1.2 Iron Gall Ink

Ten recipes were found on how to prepare iron-gall ink. Three are extracted from the treatise *Bayān al- Ṣanā ʻāt* by Teflisi (12th C), one from the treatise *Resāle-ye Buqalamun (Turkey treatise)* by Bahmani (15–16th C), four from the treatise *Majmu ʻ al-Ṣanāye ʻ (Collected Crafts)* (16th C) by Anonymous, one from *Resāle dar Bayān-e Tariqe-ye Sāḵtan-e Morakkab-e Alvān va Kāġaḏhā-ye Alvān (Treatise on Methods of Creating Coloured Paper and Ink)* (16th C) by Anonymous, and a final recipe from *Resāle dar Bayān-e, Kāġaḏ, Morakkab, va Hall-e Alvān (Treatise on Paper, Ink and Creating Colours)* (16th C) by Anonymousas seen below. As a generalisation the method involves adding vitriol (metal sulphate), to the extract of gallnuts, and sometimes, gum Arabic was used as a binder (Fig. 4.2; Table 4.3).

R1—Recipe from the treatise *Bayān al- Ṣanā ʻāt* by Teflisi (12th C):

آمیختن حبر: بستاند مازوی سپید و خرد کند و سه رطل آب بر وی ریزد. پس اندر پاتیله پاکیزه وی را در کند و نرم نرم همی جوشاند تا آبش به نیمه باز آید. پس بپالاید و رها کند سرد شود و زاگ سرخ بر وی افکند چندان که رنگش سیاه شود و بر هر رطلی آب یک درم سنگ صمغ عربی کوفته بیامیزد تا روشن آید.

Making iron-gall ink (*Hebr*): Gather white gallnuts, break to small pieces, and pour 3 *Raṭl* (1169.28 g) water on top of them in a clean pot, and boil it steadily on a fire until it reaches half its weight. Next filter, allow to cool, and add red vitriol until the colour turns black. For each *Raṭl* (389.76 g) of water, add a *Deram* (15 g) of gum Arabic to make it lighter.

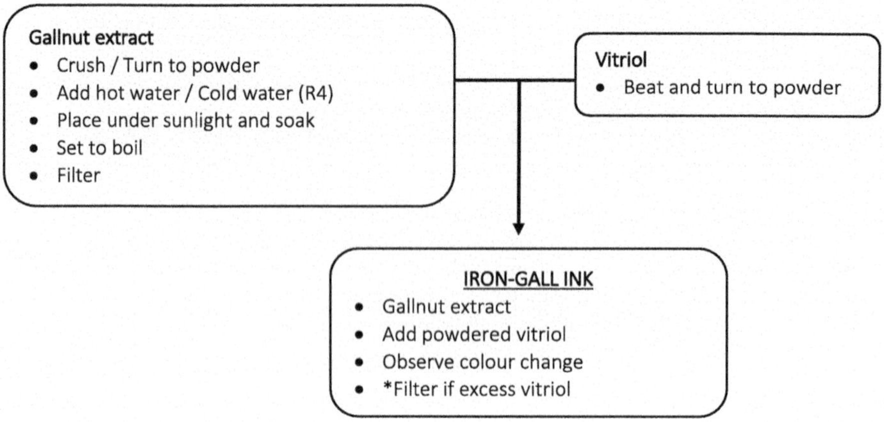

Fig. 4.2 Flowchart guide for constructing iron-gall ink

R2—Recipe from the treatise *Bayān al- Ṣanā ʻāt* by Teflisi (12th C):

بستاند مازو چندانکه خواهد و همسنگ او هشت بار آب در وی کند و همی جوشاند تا به چهار یک باز آید. پس از آن از آتش فرو گیرد و بپالاید پاکیزه، و زاگ سرخ یا زرد خرد بساید و بر وی افکند. پس روزی چند وی را به آفتاب بنهد تا نیکو و پاکیزه گردد. و اگر خواهد که روشن و تابنده آید صمغ عربی در وی افکند تا بغایت خوب بیاید.

Another method: Gather desired number of gallnuts and pour water on top, 8 times its weight, and boil until reduced to a fourth of its weight. Afterwards, remove from fire and filter it well, and add grounded red or yellow vitriol. Everyday leave it under sunlight until it is ready. If you want the ink to be lighter and lustrous, add gum Arabic.

R3—Recipe from the treatise *Bayān al- Ṣanā ʻāt* by Teflisi (12th C):

اگر خواهد که حبر ساعتی کند پنج ستیر مازو بستاند و خردش بکوبد. پس پنج درم زاگ سرخ کوفته با وی بیامیزد و مقدار یک رطل آب گرم بر وی ریزد و به کفچه نیکش بجنباند تا ممزوج شود. پس ازان وی را بپالاید و در آبگینه کند و بساعت به کار برد.

Another method: If you want to make iron-gall ink (*Hebr*) in an hour, acquire 5 *Sirs* (375 g) gall nuts and beat them well. Afterwards add to it 5 *Derams* (75 g) red vitriol and a *Raṭl* (389.76 g) of water mix well with a spatula. Afterwards filter, place in a glass container and use as desired.

R4—Recipe from the treatise *Resāle-ye Buqalamun* by Bahmani (15th C):

در ساختن مداد دیگر بستانند زاج سبز و زاج سفید و صمغ عربی برابر او و بکوبند. همه با آب مازو جوشانیده استعمال کنند.

Other ink: acquire green and white vitriol and gum Arabic equal to the weight of the vitriols and beat to powder. Boil all in gall nut extract and use as desired.

Table 4.3 Iron-gall ink—ingredients and measurements

Iron-gall ink	Primary ingredients	Parts	Amount	Secondary ingredients	Parts	Amount
R1 (12th C)	Gallnuts	–	–	–	–	–
	Red vitriol	–	–	–	–	–
	Gum Arabic	–	–	–	–	–
	Water	–	1169.3 ml	–	–	–
R2 (12th C)	Gallnuts	1 part	–	Gum Arabic	–	–
	Red/yellow	–	–	–	–	–
	vitriol	8 part	–	–	–	–
	Water					
R3 (12th C)	Gallnuts	375 g	–	–	–	–
	Red vitriol	75 g	–	–	–	–
	Water	389.76 g	–	–	–	–
R4 (15th C)	Green vitriol	1	–	–	–	–
	White vitriol	1	–	–	–	–
	Gum Arabic	2	–	–	–	–
	Gall nut	–	–	–	–	–
	extract					
R5 (16th C)	Turkish vitriol	–	–	–	–	–
	Gallnuts	–	3 kg	–	–	–
	Water	–	6 kg/L	–	–	–
R6 (16th C)	Red vitriol	–	–	–	–	–
	White gallnuts	30 pieces	–	–	–	–
	Gum Arabic	–	500 g	–	–	–
	Water	–	1.16 kg	–	–	–
R7 (16th C)	Red/yellow	–	–	–	–	–
	Vitriol	A Fourth	–	–	–	–
	Gallnuts	Half weight	500 g	–	–	–
	Gum Arabic	–	–	–	–	–
	Water					
R8 (16th C)	Red/yellow	–	30 g	–	–	–
	vitriol	–	375 g	–	–	–
	Gallnuts	–	389.76 g	–	–	–
	Water					
R9 (16th C)	Kermani	–	–	–	–	–
	vitriol	–	–	–	–	–
	Gallnuts	–	–	–	–	–
	Water					
R10 (16th C)	Vitriol	–	–	–	–	–
	Gallnuts	–	–	–	–	–
	Water	–	–	–	–	–

R5—Recipe from the treatise *Majmu ʿ al-Ṣanāye ʿ* by Anonymous(16th C):

مازوی سبز چندان که خواهند نیم کوفته کنند و مقدار یک من در دو من آب کنند و در آفتاب نهند با آب، به مقدار یک من و ده سیر باز آید و می آرند نا چنین شود که از کاغذ نگذرد اگر بگذرد بگذار تا سطبر شود وانگاه آب از سر وی برگیرند و قطره قطره آب زاگ ترکی می افزایند یعنی می چکانند تا سیاه شود و به چوبکی می جنبانند تا سیاه شود، بعده اگر آب زاگ زیاده کنند تباه گردد و به پرده از آن نگاه دارد.

Find green gallnuts as much as desired and beat gently and add 1 *Mann* (3 kg) of it to 2 *Manns* (6 kg/litre) of water and place under sunlight until 1 *Mann* and 10 *Sir* (3750 g) are left. Test it on paper so it does not penetrate to the other side. If it passes through, allow it to stay under sunlight and get more viscous. Now add Turkish vitriol water in drops until the liquid darkens. Mix with a stick and it shall become black. If too much vitriol was used, the ink will decay, so filter it through cloth to remove the excess vitriol.

R6—Recipe from the treatise *Majmu‘ al-Ṣanāye‘* by Anonymous (16th C):

بر لب لطیف سی عدد مازوی سپید را خرد کنند و سه رطل آب در وی ریزند و در پاتیله نرم می جوشانند تا آبش به نیم آید. پس بپالایند و رها کنند تا سرد شود، و زاگ سرخ در وی ریزند و صلایه کنند چنانچه در هر سیر و نیم آب، یک درم صمغ عربی کوفته، بیندازند تا روشن شود، بعده بکار برند.

Bray 30 pieces of white gallnuts and pour 3 *Raṭl* (1.16 kg) of water over it. Set to a gentle boil in a copper pan until the water has evaporated to half its amount. Filter the liquid and let it cool down. Add red vitriol and bray in the manner that in every 1 ½ *Sirs* of water (112.5 ml), 1 *Deram* (3.2 g) of brayed gum Arabic is added to make it brighter1 and use as desired.

R7—Recipe from the treatise *Majmu‘ al-Ṣanāye‘* by Anonymous (16th C):

بستاند مازو چهار یکی، و نیم سنگ صمغ عربی در وی کنند و همی جوشانند تا به چهار یک باز آید، آنگاه فرو گیرند و بپالایند و زاج سرخ یا زردی در وی افکنند، ساویده. پس او را روزی چند در آفتاب نهند تا نیکو شود و پاکیزه گردد و اگر خواهند که روشن شود صمغ عربی آمیزند و به کار برند به غایت نیک آید.

Acquire a fourth of gallnuts and add half weight (*Sang*) of gum Arabic and set to boil until it is down to a fourth of its weight. Then filter and add red or yellow vitriol and set to bray. Leave it out under the sun for a few days until it becomes nice and clean. Should you desire to make a brighter ink, add more gum Arabic and use, for it shall be a beautiful ink.

R8—Recipe from the treatise *Majmu‘ al-Ṣanāye‘* by Anonymous (16th C):

پنج سیر مازو بستانند فرو کوبند و پنج درم زاک کوفته با وی بیامیزند و مقدار یک رطل بر وی ریزند و به کفچه نیک بجنبانند ممزوج شود آنگاه بپالایند و در ظرف آبگینه کنند و بکار برند.

Acquire 5 *Sirs* (375 g) of gall and beat well and add to it 5 *Derams* (30 g) of ground vitriol and a *Raṭl* (389.76 g) of water. Mix the ink well then filter through cloth. Finally, pour the said ink it into a glass bottle, and use as necessary.

R9—*Resāle dar bayān-e Tariqe-ye Sāḵtan-e Morakkab-e Alvān va Kağaḏhā-ye Alvān* by Anonymous (16th C):

قدرو مازوی رسیده را خرد بساید و در آب کند، پنج روز بعد از آن در هرکاره سنگین چندین بجوشاند تا بقوام آید. بعد از آن قدری زاج کرمانی سوخته را در آب مازو کند و به هم بیامیزد تا سیاه گردد. آن را به رگوی محکم بیالاید و بدان کتابت کند اما نوشته را از نم نگاه دارد تا نشو نکند.

Bray fresh gall, add it to water and after 5 days, set it to boil in a stone pot. Then add some burnt ‘Kermani’ (city in Iran) vitriol in the gall water and stir. Once the liquid darkens, filter it through a fine cloth. You may now start writing but make sure the paper is kept in place free of humidity, so the ink will not bleed.

R10—Recipe from the treatise *Resāle dar Bayān-e, Kağad, Morakkab, va Hall-e Alvān* by Anonymous (16th C):

قدری مازو را خرد کند و سه روز در آب بگذارد. پس آن آب صاف را در هرکاره سنگین و آتش نرم نرم کند و چنان به قوام آید که چون بر کاغذ نویسند، نشو نکند. آنگاه زاگ پاک بی گوگرد و خاک با وی بیامیزد و بپالاید. مدادی نیکو است اما از نم نگاه باید داشت، که نشو نکند که اوراق بریکدیگر چسبد.

> Break some galls and place in water for 3 days. Filter the liquid and pour it into a stone pot and boil it gently to the point that if a drop of it is placed on paper, it will not pass through the reverse side. In the next step, add clean vitriol free of dust and sulphur, mix together and filter this liquid. It will be a good ink, but make sure the paper is free of moisture, so the ink will not pass through, resulting in the papers sticking to each other.

Iron-gall ink is the product of a chemical reaction between a vitriol (iron sulphate) and tannic acid extracted from gallnuts. It was the most widely used writing material in western history (Hahn 2010, p. 44). It is more permanent than carbon ink, due to its ability to penetrate paper fibres and being less water-soluble (Carvalho 1904). However, iron gall ink was known to degrade paper. This is caused by a combination of: acid hydrolysis (due to the presence of sulphuric acid) and oxidation of the cellulose in paper catalysed by Fe^{2+} ions in the ink (especially if there is an excess of ferrous sulphate), and other transition metals. Acids weaken paper, migrating outwards from the ink resulting in halos around the text, as well as migrating to neighbouring pages, in the process decreasing the visibility of the ink (Botti et al. 2005). A recipe on creating iron-gall ink can be found in Fig. 4.3.

4.1.3 Persian Ink (Carbon-Tannate Ink)

Persian ink is a mixture of carbon ink and iron-gall ink in set proportions. This ink has more fluidity and darkness compared to carbon ink, with the added benefit of penetrating paper more deeply than carbon ink. As a result, it is more permanent than carbon ink, whilst being less detrimental to paper compared to iron-gall ink. Eighteen recipes were found on the preparation of this ink: two from the treatise *Ṣerāṭ al-Soṭur* (Mašhadi, 15th C), two from *Resāle dar Bayān-e, Kağad, Morakkab, va Hall-e Alvān* (Anonymous 16th C), three from *Medād al-Ḵoṭuṭ* (Heravi 16th C), one from *Rasm al-Ḵaṭ* (Heravi 16th C), two from *Savād al-Ḵaṭ* (Heravi 16th C), one from *Adāb al-Mašq* (Heravi, 16th C), three from *Majmu ʻ al-Ṣanāye ʻ* (Anonymous 16th C), two from the treatises *Resāle dar Bayān-e Ḵaṭ va Morakkab va Kağad va Sāḵtan-e Ranghā* (Anynymous 15th C) one from the treatise *Resāle dar Bayān-e Tariqe-ye Sāḵtan-e Morakkab-e Alvān va Kağadhā-ye Alvān* (Anonymous 16th C), and a final recipe from the treatise *Golestān-e Honar* by Qāżi Aḥmad Qomi (16th C) (Fig. 4.4; Table 4.4).

Poem-recipe:

همسنگ دوده زاج است همسنگ هر دو مازو　　همسنگ هر سه صمغ است آنگاه زور بازو

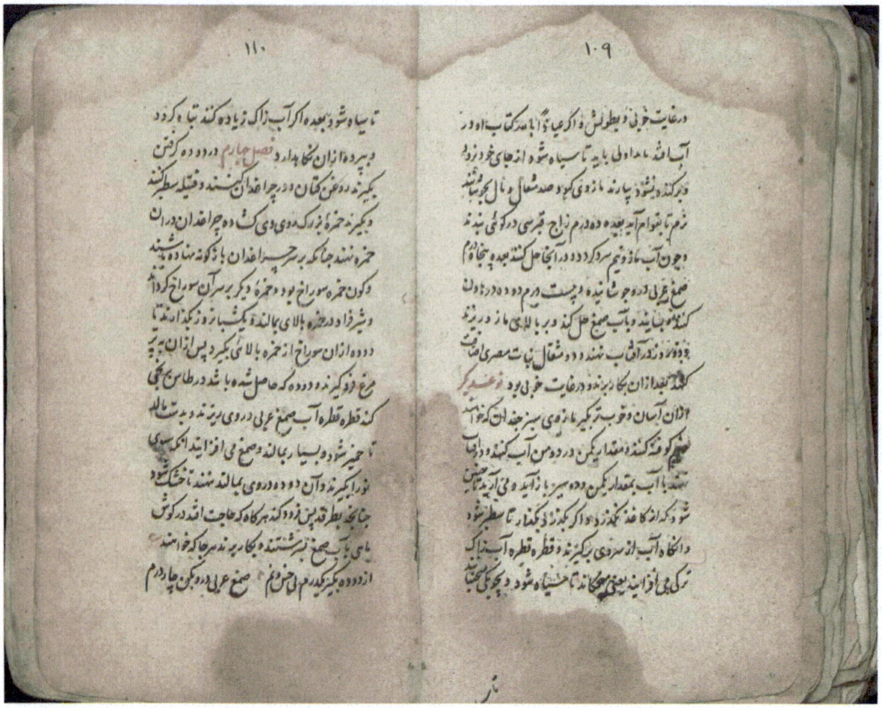

Fig. 4.3 Guidelines on preparing lamp black (left page) and recipe on Iron-gall ink (right page) from the treatise Majmu 'al-Ṣanāye' (16th C), Manuscripts 3875—Tehran University Central Library

The equal of soot is vitriol, the equal of both combined is gall.

The equal of all three is gum, and so, the power of your muscles.

The recipe-poem gives the ingredients needed to create this ink, with its exact proportions: one-part soot, one part vitriol, two parts gallnuts and four parts gum Arabic. The power of muscle relates to the act of braying and mixing the ink (Zekrgoo 2014, p. 137). This poem can be seen many treatises, and the same poetry is often used in recipes on creating Peacock (*Moṭavvas*) ink, a variation of the carbon-tannate ink with additional secondary ingredients, which is the focus of the next ink category.

R1—Poem-recipe by Ibn Muqla' Farsi (d.930 C) found in the treatise *Medād al-Ḵoṭuṭ* by Mir 'Ali Heravi (16th C):

بستان دو درم دود چراغ بی نم   صمغ عربی فکن در او چار درم

مازو دو درم نیم درم زاج سفید   از بهر مرکبش فروسای بهم

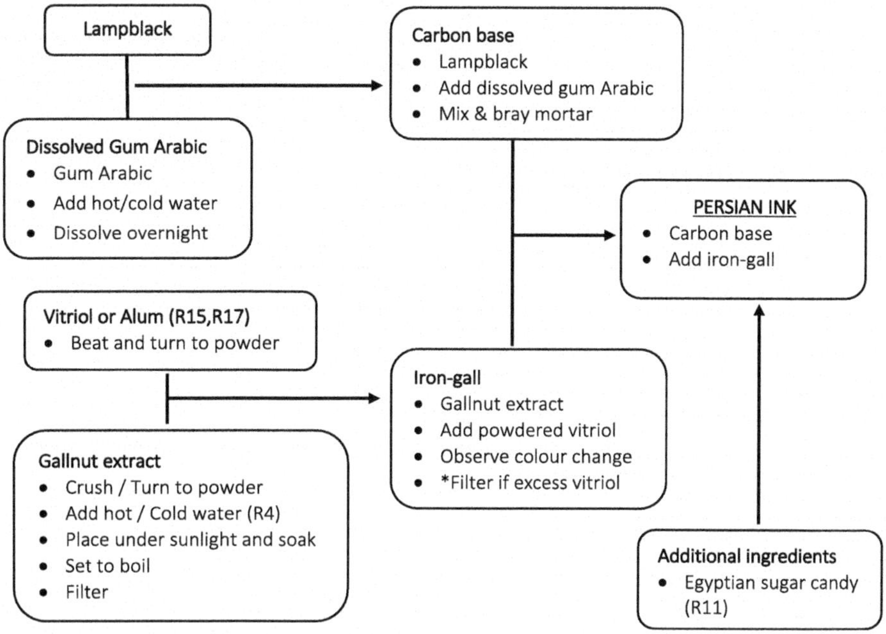

Fig. 4.4 Flowchart guide for constructing Persian ink

Get two Derams (30 g) of dry soot; add four Derams (60 g) of gum Arabic. Two Derams (30 g) of gallnuts, half Deram (7.5 g) of white vitriol; to make the ink mix and bray them together.

R2—Poem Recipe from the treatise *Resāle dar Bayān-e, Kağad, Morakkab, va Hall-e Alvān* by (Anonymous 15th C)

همسنگ دوده زاج است همسنگ هر دو مازو   همسنگ هر سه صمغ است آنگاه زور بازو

'The equal of soot is vitriol, the equal of both combined is gall.
The equal of all three is gum, and so, the power of your muscles'.

R3 Poem-recipe from the treatise *Resāle dar Bayān-e, Kağad, Morakkab, va Hall-e Alvān* (Anonymous 15th C):

بستان دو درم دود چراغ بی نم   صمغ عربی فکن در او چار درم
مازو سه درم نیم درم ترکی زاج   از بهر مرکبش فروسای بهم

Get two Derams (30 g) of dry soot; add four Derams (60 g) of gum Arabic. Three Derams of gallnuts (75 g), half Deram of Turkish vitriol (7.5 g); to make the ink mix and bray them together.

Table 4.4 Carbon-tannate (Persian) Ink—ingredients and measurements

Persian ink	Primary ingredients	Parts	Amount	Secondary ingredients	Parts	Amount
R1 (10th C)	Lamp black	–	30 g	–	–	–
	White vitriol	–	1.6 g	–	–	–
	Gallnuts	–	30 g	–	–	–
	Gum Arabic	–	7.5 g	–	–	–
	Water	–	–	–	–	–
R2 (15th C)	Lamp black	1	–	–	–	–
	Vitriol	1	–	–	–	–
	Gallnuts	2	–	–	–	–
	Gum Arabic	4	–	–	–	–
	Water	–	–	–	–	–
R3 (15th C)	Lamp black	–	30 g	–	–	–
	Turkish vitriol	–	7.5 g	–	–	–
	Gallnuts	–	75 g	–	–	–
	Gum Arabic	–	60 g	–	–	–
	Water	–	–	–	–	–
R4 (15th C)	Lamp black	–	75 g	–	–	–
	Vitriol	–	75 g	–	–	–
	Gallnuts	–	150–225 g	–	–	–
	Gum Arabic	–	300 g	–	–	–
	Water	–	–	–	–	–
R5 (15th C)	Lamp black	–	75 g	–	–	–
	Alum	–	75 g	–	–	–
	Gallnuts	–	150–225 g	–	–	–
	Gum Arabic	–	300 g	–	–	–
	Water	–	–	–	–	–
R6 (15th C)	Lamp black	–	75 g	–	–	–
	Vitriol	–	75 g	–	–	–
	Gallnuts	–	150–225 g	–	–	–
	Gum Arabic	–	300 g	–	–	–
	Water	–	–	–	–	–
R7 (15th C)	Lamp black	–	75 g	–	–	–
	Alum	–	75 g	–	–	–
	Gallnuts	–	150–225 g	–	–	–
	Gum Arabic	–	300g	–	–	–
	Water	–	–	–	–	–
R8 (16th C)	Lamp black	1	–	–	–	–
	Vitriol	1	–	–	–	–
	Gallnuts	2	–	–	–	–
	Gum Arabic	4	–	–	–	–
	Water	–	–	–	–	–
R9 (16th C)	Lamp black	1	–	–	–	–
	Vitriol	1	–	–	–	–
	Gallnuts	2	–	–	–	–
	Gum Arabic	4	–	–	–	–
	Water	–	–	–	–	–

(continued)

Table 4.4 (continued)

Persian ink	Primary ingredients	Parts	Amount	Secondary ingredients	Parts	Amount
R10 (16th C)	Lamp black	1	–	–	–	–
	Vitriol	1	–	–	–	–
	Gallnuts	2	–	–	–	–
	Gum Arabic	4	–	–	–	–
	Water	–	–	–	–	–
R11 (16th C)	Lamp black	–	15 g	–	–	–
	Kotan vitriol	–	7.5 g	–	–	–
	Gallnuts	–	30 g	–	–	–
	Gum Arabic	–	60 g	–	–	–
	Water	–	–	–	–	–
R12 (16th C)	Lamp black	1	–	–	–	–
	Vitriol	1	–	–	–	–
	Gallnuts	2	–	–	–	–
	Gum Arabic	4	–	–	–	–
	Water	–	–	–	–	–
R13 (16th C)	Lamp black	1	–	–	–	–
	Vitriol	1	–	–	–	–
	Gallnuts	2	–	–	–	–
	Gum Arabic	4	–	–	–	–
	Water	–	–	–	–	–
R14 (16th C)	Lamp black	1	–	–	–	–
	Vitriol	1	–	–	–	–
	Gallnuts	2	–	–	–	–
	Gum Arabic	4	–	–	–	–
	Water	–	–	–	–	–
R15 (16th C)	Lamp black	–	300 g	Egyptian sugar candy	9.28 g	–
	Cyprus Vitriol	–	150 g		–	–
	Gallnuts	–	464 g	–	–	–
	Gum Arabic	–	750 g	–	–	–
	Water	–	–	–	–	–
				–		
R16 (16th C)	Lamp Black	–	–	–	–	–
	Vitriol	–	–	–	–	–
	Gallnuts	1	–	–	–	–
	Gum Arabic	1	–	–	–	–
	Water	12	–	–	–	–
R17 (16th C)	Lamp black	–	–	–	–	–
	Vitriol	–	–	–	–	–
	Gallnuts	–	9.28 g	–	–	–
	Gum Arabic	–	18.56 g	–	–	–
	Water	–	–	–	–	–
R18 (16th C)	Lamp black	1	–	–	–	–
	Vitriol	1	–	–	–	–
	Gallnuts	2	–	–	–	–
	Gum Arabic	4	–	–	–	–
	Water	–	–	–	–	–

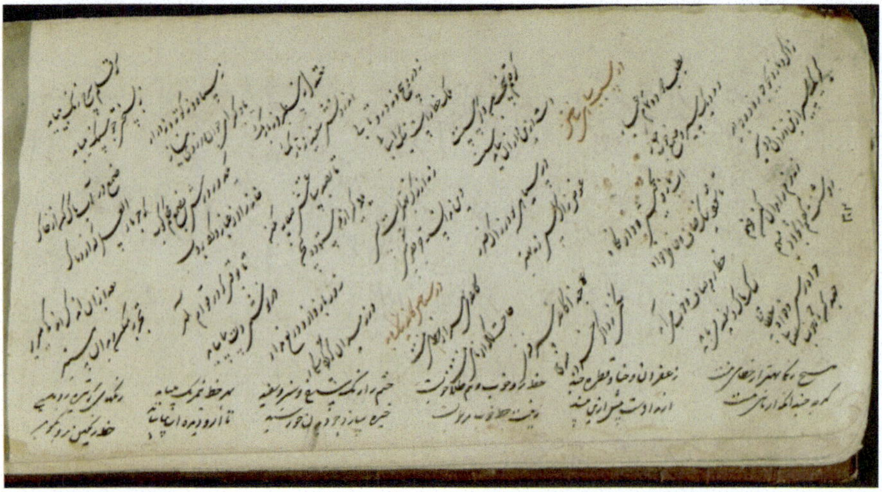

Fig. 4.5 Poem Recipe from the treatise Ṣerāṭ al-Soṭur by Solṭān ʻAli Mašhadi (15th C) on making Persian ink and using alum instead of vitriol. Image courtesy of Tehran University Central Library (manuscript 4736)

In a single poem, Mašhadi (15th C) gives a recipe on making Persian ink, where he also mentioned that it is best for vitriol to be replaced by alum, hence giving 2 recipes (R4 & R5). The exact same poem is seen in another treatises (R6 & R7):

R4 & R5—Poem Recipe from the treatise *Ṣerāṭ al-Soṭur* by Solṭān ʻAli Mašhadi (15th C) (Fig. 4.5):

R6 & R7—Poem Recipe from the treatise *Resāle dar bayān-e Ḵaṭ va Morakkab va Kağaḏ va Sāḵtan-e Ranghā* by Anonymous (15th C):

بطلب دوده تمام عیار — دوده یک سیر و صمغ خوی چهار
زاج و مازو بجو چه زود و چه دیر — گیر یک سیر ازین و زان دو سه سیر
صمغ در آب ریز پاک زخاک — تا چو ماء العسل گدازد پاک
یک دو روزش به صمغ محکم کوب — خانه را از غبار پاک بروب
تا به صد ساعتش صلایع بکن — یاد گیر از من این ستوده سخن
زمه از زاک بهتر است بسی — وین ندانسته جز فقیر کسی
در سیاهی بود ززاک ضرر — عوض زاگ بس زمه بهتر
آب مازو بجوش و دار نگاه — تا شود نیک صاف و خاطر خواه
زمه نرم را بدو کن ضم — روشنت گفتم آنچه بود مبهم
بعد از آن اندک اندکش میریز — تجربه می کن و بدو مستیز
تا به وقطی که به قوام آید — در نوشتن دلت بیاساید
زور بازو ازو دریغ مدار — ورنه می دان که کرده ای بیگار

Find some soot of the highest quality, one Sir (75 g) of soot and four Sirs (300 g) of good gum.

Find vitriol and gall, a Sir (75 g) of former and two (150 g) or three (225 g) of the latter.

Pour the gum in dust free water until it's clear and takes the form of honey.

Bray this gum water for two days, clear the house from any dust.

Mix and bray the ink for a hundred hours and learn these words of wisdom from me

Alum (Zame) is much better than Vitriol (Zāg); only this humble one acquired this knowledge.

Vitriol can damage the ink hence, it's better to use Alum instead.

Boil the essence of gall and keep it until it becomes clear, pure, and desirable. Now add the beaten alum, I have enlightened you on what was unclear.

Add alum in small portions, do not hesitate and test it.

Until the time when it settles, and your heart feels at ease when writing. Do not spare the strength of muscle otherwise your work was in vain.

R8—Poem Recipe from the treatise *Medād al-Ḵoṭuṭ* (Heravi 16th C)

همسنگ دوده زاج است همسنگ هر دو مازو   همسنگ هر سه صمغ است آنگاه زور بازو

The equal of soot is vitriol, the equal of both combined is gall.

The equal of all three is gum, and so, the power of your muscles

R9—Poem Recipe from the treatise *Savād al-Ḵaṭ* (Heravi 16th C)

همسنگ دوده زاج است همسنگ هر دو مازو   همسنگ هر سه صمغ است آنگاه زور بازو

R10—Poem Recipe from the treatise *Resāle dar Bayān-e Tariqe-ye Sāḵtan-e Morakkab-e Alvān va Kağaḏhā-ye Alvān* (Anonymous 16th C).

همسنگ دوده زاج است همسنگ هر دو مازو   همسنگ هر سه صمغ است آنگاه زور بازو

The equal of soot is vitriol, the equal of both combined is gall.

The equal of all three is gum, and so, the power of your muscles

R11—Poem-recipe from the treatise *Majmu‘ al-Ṣanāye‘* (Anonymous 16th C):

از دوده بگیر یک درم بی خس و بی غم   صمغ عربی در او بکن چار درم
مازو دو درم نیم درم زاگ ختن   از بهر مرکبش فروسای بهم

Get a Deram (15 g) of dry soot and add four Derams (60 g) of Gum Arabic.

Two Derams (30 g) of gall, half Derams of Ḵotan[1] vitriol (7.5 g); to make the ink mix and bray them together.

R12—Poem-recipe from the treatise *Medād al-Ḵoṭuṭ* by Mir ‘Ali Heravi (16th C):

چنین گفت آنکه او اهل تمیز است   که اجزای سیاهی چار چیز است
نخستین دوده پاک و دوم زاج   ولی همسنگ باید دوده را زاج
سیم جزو سیاهی هست مازو   ولی مازو بود همسنگ هر دو
بود جزو چهارم صمغ خوشرنگ   ولی با هر سه جزوش ساز همسنگ
اگر خواهی که بشناسی سیاهی   بگویم با تو تحقیقش کماهی
سیاهی نزد ما نیکوتر آن است   که اندر دیده براق و روان است
ولیکن چون سواد دیده باید   که در آبش خلل دیده نیاید

These are the words of one who is of purity: there are four elements to ink.

First is pure soot and second is vitriol but soot should be of equal weight to vitriol.
The third element of ink is gallnuts, but its weight should be equal to the two combined.
Good coloured gum is the fourth element but make its weight equal to the rest combined.
If you want to be an expert on ink, I shall impart you this knowledge.
To us, a good ink is one that is shiny and fluent when gazed upon.
However, the liquid should be free from impurity, just like looking at the iris of the eye.

R13—Poem-recipe found in the treatise *Rasm al-Ḵaṭ* by Majnun Rafiqi Heravi (16th C) (Fig. 4.6):

چنین گفت آنکه او اهل تمیز است   که اجزای سیاهی چار چیز است
نخستین زاک و دوم دوده پاک   ولی همسنگ باید دوده با زاک
سیم جزو سیاهی هست مازو   ولی مازو بود همسنگ هر دو
بود جزو چهارم سنگ خوشرنگ   ولی با هر سه جزوش ساز همسنگ
سیاهی چون به پالودش رسانی   زکرباس لطیفش بگذرانی

[1] A town in Xinjiang province of China

Fig. 4.6 Poem recipe on making Persian ink from the treatise Rasm al-Ḵaṭ by Majnun Rafiqi Heravi (Manuscript 3522), Image courtesy of Tehran University Central Library

These are the words of one who is of purity: there are four elements to ink.

First is vitriol and second is pure soot but soot should be of equal weight to vitriol.
The third element of ink is gallnuts, but its weight should be equal to the two combined.
Good coloured gum is the fourth element but make its weight equal to the rest combined.
Once you finished braying your ink, pass your ink through a fine tent cloth.

R14—Poem-recipe from the treatise *Ādāb al-Mašq* by Majnun Rafiqi Heravi (16th C):

تعریف سیاهی و مرکب   ممن به تجرب المجرب
باشد سیه و روان و براق   اما چو سواد چشم عشاق
وز آب نگیرد الخلالی   وزنم نپرید الحلالی
ترکیب مرکبی که نیکوست   از دوده و صمغ و زاج و مازوست
دوده است به زاج همترازو   همسنگ به زاج و دوده مازو
جزو دگر است صمغ خوشرنگ   صمغ است ولی بهر سه همسنگ

To know and understand good ink, you need to have a lot of experience.

It must be black, fluent, and shiny, just like the darkness of the lover's eyes.

The water used should not have impurity and it (ink) should not deteriorate with moisture.

The best recipe for an elegant ink is lamp black, gum, vitriol, and gal-nuts.

Soot and vitriol of the same portion, but gall-nuts equal to the portion of both combined.

Another element is some good, coloured gum however equal to portion of all three combined.

And he follows with the following poem on all the steps, which need to be taken, from preparing the lamp black, to filtering the ink at the final stage.

خواهی که در آن بود سیاهی   از دوده بگیر آنچه خواهی
کاغذ طلب و خریطه ای ساز   وان دوده در آن خریطه انداز
آن را به خمیر گیر چون نان   در گرم تنور پخته گردان
گرمی بدر آر از آن فطیرش   در صمغ زلال کن خمیرش
زان پیش که طبخ دوده باشد   صمغ تو در آب بوده باشد
آبی که نتیجه حیات است   صمغی که نمونه نبات است
چون صمغ در آب او بکن جهد   چندان که نبات او شود شهد
وان دوده صمغ را به خوبی   در هاون روحیی بکوبی
چون تیره شود در آب مازو   در وی بکن آب زاج و مازو
هر گه که تمام گشت مایه   یک هفته مباش بی صلایه

If you want to create and use ink, get as much lamp black as you desire. Make a container with paper and place the lamp black in that container. Secure it in dough and bake it in a heated furnace as you do with bread. Once the dough rises due to heat remove the soot and bray it with clear gum water. Remember that while the dough was baking the gum had been dissolved in water; water that is the result of nature, gum that looks like rock sugar. Leave the gum in the water until the liquid takes the form of syrup. The lamp black and gum water should be finely brayed and mixed in a mortar. Once the gall nut extract is darkened, add it along with vitriol to the soot. Once there is no extra liquid (due to bray) make sure to bray it for a week

R15—Recipe from the treatise *Majmu' al-Ṣanāye'* by Anonymous (16th C):

بیارند مازوی کبود صد مثقال و بجوشانند نرم،تا به قوام آید، بعده ده درم زاج قبرسی در رکویی بندند و چون آب مازو نیم سرد گردد در آنجا حل کنند، بعده پنجاه درم صمغ عربی در او جوشانیده و بیست درم در هاون کنند و بسایند و به آب صمغ حل کنند و بر بالای مازو ریزند و در روز در آفتاب نهند و دو مثقال نبات مصری اضافت کنند، بعد از آن بکار برند در غایت خوبی بود.

Bring 100 *Meṯqāls* (464 g) of dark gallnuts and boil gently in water until it reaches the desired state. Secure 10 *Derams* (150 g) of Cyprus vitriol in a piece of cloth and once the gall nut extract is half cooled, place the vitriol bag in and allow it to dissolve. Meanwhile boil 50 *Derams* (750 g) of Gum Arabic and bray and beat 20 *Derams* (300 g) of soot in a mortar. Mix the soot and gum water well and pour this mixture on top of the gall-vitriol mixture and place under sunlight during the day. Then add 2 *Meṯqāls* (9.28 g) of Egyptian sugar candy and use it for it shall be a good ink.

R16—Recipe from the treatise *Majmu' al-Ṣanāye'* by Anonymous (16th C):

مازوی سبز بی سوراخ یک جزو در دوازده جزو آب آغارند و یک شبانروز بگذارند و بجوشانند تا شش جزو باز آید و یک جزو صمغ عربی پاکیزه در آب آغارند چنانکه آن را نیم کند، آنگاه صافی کنند و هر دو آب را بپالایند و چون سرد شود به هم بیامیزند چنانکه یک بنفش رنگ گردد و باز بپالایند تا به وقت حاجت با دوده نیک بیامیزند و صلایه کنند به غایت نیک آید.

Place one-part green gallnuts without holes in twelve parts of water, and leave overnight, then boil it until it the water loses half its volume (six parts). Now add one part gum Arabic to two parts of water. Filter them both and when they are cooled down, mix, and bray them together until it turns purple. Then filter this ink and whenever needed, mix it with lamp black and set to bray for it shall result in a fine ink.

R17—Recipe from the treatise *Savād al-Ḵaṭ* by Majnun Rafiqi Heravi (16th C):

چهار مثقال صمغ و دو مثقال مازوی سوده با دوده بیامیزند و آب گرم در وی ریزند و صلایه کنند و زاک در میان کاغذ کرده در خمیر گیرند و در زیر آتش کرده تا بپزد، آنگاه بسایند و با دوده بیامیزند و مقداری بجوشانند و از کرباس سطبر بگذرانند. پس بر بالای غربال خشک نمایند، و هر گاه که خواهند بدان کتابت نمایند. مقداری در کاسه ریزند و قدری آب گرم در وی کنند و یک شب بگذارند، مرکبی در غایت لطافت باشد.

Mix 4 *Meṯqāl*s (18.56 g) of gum and 2 *Meṯqāl*s (9.28 g) of brayed gallnuts with lamp black in warm water and beat them well in a mortar. Place some vitriol in a paper container, secure it in dough and bake it in a furnace. Afterwards beat the heated vitriol and mix it with the other ingredients and set to bray. In the next step the mixture is set to a gentle boil and filtered through cloth. Now let this ink dry on a screen and whenever needed, place some in a bowl, add hot water and let it soak overnight. It shall be a smooth ink.

R18 Poem-recipe from the treatise *Golestān-e Honar* by Qāżi Aḥmad Qomi (16th C):

همسنگ دوده زاج است همسنگ هر دو مازو   هم وزن هر سه صمغ است آنگاه زور بازو

The equal of soot is vitriol, the equal of both combined is gall.

The equal of all three is gum, and so, the power of your muscles.

4.1.4 Peacock (Moṭavvas/Ṭavoosi) Ink

A type of ink that was celebrated mostly by the Persians for its quality, colour and its permanence, was called *Ṭavusi* or *Moṭavvas* which means peacock or peacock-like. The ink gets its name due to the variety of different coloured materials used in its production as well as the blue-green undertone of the predominantly black ink. It is a variation of the carbon-tannate ink category with regard to its primary ingredients having similar proportions.

However, a variety of secondary ingredients were added by different master calligraphers experimenting with perfecting the ink. Of all inks, peacock ink was most valued by master calligraphers. A total of 18 recipes were found on this ink. For information regarding the secondary ingredients and the reason for its usage, refer to 'Chapter Four: Constituents of Persian Inks'(Fig. 4.7 and Table 4.5).

R1—Recipe-poem by K̲āje Jamāl al-Ddin Yāqut (d. 1298) found in the treatise *Medād al-K̲oṭuṭ* by Mir 'Ali Heravi (16th C):

ای که در علم خط شدی استاد   من مدادی ترا بیاموزم
زاج اسپید و مازوی اخضر   دوده بذر بیغش و بی نم
صمغ برالحجاز و ماء الورد   مشک خالص نبات مصری هم
وسمه و مورد را عصاره بگیر   اندکی زعفران کنش منظم
ورق زر و نقره داخل کن   تا چهل روز سهق کن محکم
وزن هر یک به تجربت حاصل   می شود بی گمان نه بیش و نه کم
چون محک سازیش نماید زرد   نیست مصلش به عرصه عالم

Oh, you who have mastered the art of scripts, I shall educate you on a type of ink

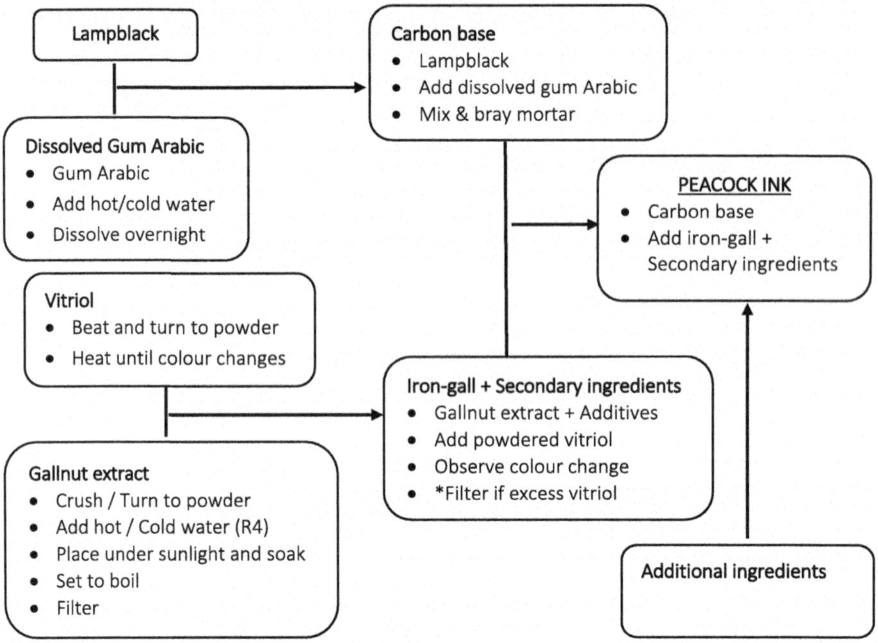

Fig. 4.7 Flowchart guide for constructing Peacock ink

Table 4.5 Peacock ink—ingredients and measurements

Peacock ink	Primary ingredients	Parts	Amount	Secondary ingredients	Parts	Amount
R1 (13th C)	Lamp black	–	–	Myrtle	–	–
	White vitriol	–	–	Woad	–	–
	Green	–	–	Egyptian sugar candy	–	–
	gallnuts	–	–	Musk	–	–
	Hejaz Gum	–	–	Rose water	–	–
	Arabic			Gold	–	–
	Water			Silver	–	–
				Saffron	–	–
R2 (14th C)	Lamp black	6	–	Cypress aloe	9	–
	Gum Arabic	9	–	Rock sugar	9	–
	Water	–	–	Verdigris	2	–
				Saffron	1	–
				Rose water	–	–
				Myrtle	–	
R3 (15th C)	Lamp black	–	150 g	Henna	–	–
	Turkish vitriol	–	75 g	Myrtle	–	–
	Gallnuts	–	225 g	Woad	–	–
	Gum Arabic	–	300 g	Dogrose extract	–	–
	Water	–	–	Saffron water	–	–
				Ground aloe	–	–
				Salt	–	–
				Ground coral/pearl	–	–
				Musk	–	–
				Liquid amber	–	–
				Gold	–	–
				Silver	–	–
				Copper	–	–
				Brass	–	–
				Cinnabar	–	–
				Lapis lazuli	–	–
				Sandarac	–	–
R4 (15th C)	Lamp black	1	–	Saffron	–	4.64 g
	Cyprus vitriol	1	–	Rose water	–	150 g
	Gallnuts	2	–	Myrtle	–	15 g
	Gum Arabic	4	–	Woad	–	7.5 g
	Water	–	–	Henna	–	15 g
				Tibetan musk	–	7.5 g
				Epithymum	–	7.5 g
				Indigo	–	–
				Aloe	–	–
				Indian Salt	–	–
				Egyptian Sugar Candy	–	–
R5 (16th C)	Lamp black	1	–	Henna	–	15 g
	Vitriol	1	–	Myrtle	–	15 g
	Gallnuts	2	–	Epithymum	–	3.75 g
	Gum Arabic	4	–	Indigo	–	–
	Water	10	–	Egyptian Sugar Candy	–	–
				Salt	–	–
				Musk	–	–
				Rose Water	–	–

(continued)

Table 4.5 (continued)

Peacock ink	Primary ingredients	Parts	Amount	Secondary ingredients	Parts	Amount
R6 (16th C)	Lamp black White vitriol Gum Arabic Water	6 1 1 –	– – – –	Cypress aloe Rock sugar Saffron Indigo Verdigris Rose water Myrtle	1 1 2 2 2 –	– – – – –
R7 (16th C)	Lamp black White vitriol Gallnuts Gum Arabic Water	– – – – –	69.6 g 69.6 g 139.2 g 278.4 15 kg	Henna Myrtle Woad Egypt sugar candy Musk Saffron Indigo Epithymum Aloe Rose water	– – – – – – – – – –	75 g 75 g 75 g 15 g 7.5 g 15 15 g 15 g – –
R8 (16th C)	Lamp black Cyprus vitriol Gallnuts Gum Arabic Water	– – – – –	150 g 150 g 75 g 525 g 21 kg	Henna Myrtle Woad Saffron Egypt rock sugar Sal ammoniac	– – – – – –	75 g 75 g 75 g 15 g 15 g 7.5 g
R9 (16th C)	Lamp black Vitriol Gallnuts Gum Arabic Water	– – – – –	150 g 75 g 300 g 525 g 15L	Henna Myrtle Saffron Egypt rock sugar Sal ammoniac	– – – – –	75 g 75 g 15 g 15 g 7.5 g
R10 (16th C)	Lamp black Vitriol Gallnuts Gum Arabic Water	1 1 2 4 –	– – – – –	Camel thorn Myrtle Woad Egyptian sugar candy Saffron Epithymum Musk Rose water Aloe Indian salt Indigo	– – – – – – – – – – –	7.5 g 7.5 g 7.5 g – 4.64 g 7.5 g 7.5 g 150 g – – –
R11 (16th C)	Lamp black Turkish vitriol Gallnuts Gum Arabic Water	– – – – –	75 g 150 g 300 g 300 g	Saffron Sugar candy Aloe Rose Water	– – – –	2.32 g 37.5 g – –
R12 (16th C)	Lamp black Vitriol Gallnuts Gum Arabic Water	– – – – –	– – – – –	Henna Myrtle Woad Epithymum Green walnut husks	– – – – –	– – – – –

(continued)

Table 4.5 (continued)

Peacock ink	Primary ingredients	Parts	Amount	Secondary ingredients	Parts	Amount
R13 (16th C)	Lamp black	–	46.4 g	Saffron	–	2.32 g
	Vitriol	–	185.6 g	Epithymum	–	9.28 g
	Gallnuts	–	371.2 g	Green walnut husks	–	46.4 g
	Gum Arabic	–	185.6 g	Indigo	–	9.28 g
	Water	–	–	Aloe	–	–
R14 (16th C)	Lamp black	–	46.4 g	Aloe Vera	–	9.28 g
	Marcasite	–	23.2 g	Henna	–	1.5 kg
	Gum Arabic	–	18.56 g	Myrtle	–	1.5 kg
	Water	–	–	Rose Water	–	–
R15 (16th C)	Lamp black	–	150 g	Myrtle	–	–
	Turkish vitriol	–	75 g	Woad	–	–
	Gallnuts	–	225 g	Henna	–	–
	Gum Arabic	–	120 g	Saffron	–	–
	Water	–	–	Rose water	–	–
				Aloe	–	–
				Salt	–	–
				Coral	–	–
				Chicory extract	–	–
R16 (16th C)	Lamp black	1	–	Saffron	–	4.64 g
	Cyprus vitriol	1	–	Rose water	–	150 g
	Gallnuts	2	–	Myrtle	–	15 g
	Gum Arabic	4	–	Woad	–	7.5 g
	Water	–	–	Henna	–	15 g
				Tibetan musk	–	7.5 g
				Epithymum	–	7.5 g
				Indigo	–	–
				Aloe	–	–
				Indian salt	–	–
				Egyptian sugar candy	–	–
R17 (16th C)	Lamp black	1	–	Henna	–	15 g
	Cyprus vitriol	1	–	Myrtle	–	15 g
	Gallnuts	2	–	Epithymum	–	7.5 g
	Gum Arabic	4	–	Woad	–	7.5 g
	Water	–	–	Saffron	–	9.28 g
				Musk	–	7.5 g
				Rose Water	–	150 g
				Rock Sugar	–	–
				Indian salt	–	–
				Indigo	–	–
				Aloe	–	
R18	Lamp black	–	150 g	Henna	–	75 g
	Cyprus vitriol	–	75 g	Myrtle	–	75 g
	Gallnuts	–	300 g	Saffron	–	15 g
	Gum Arabic	–	525 g	Egyptian rock sugar	–	15 g
	Water	–	15 kg	Sal Ammoniac	–	7.5 g

White vitriol, green gallnuts and unadulterated dry soot from seeds

Hejaz Gum and rose water, pure musk and Egyptian sugar candy

Extract the essence of woad and myrtle; add a little bit of saffron

Insert leafgold and leafsilver; beat it well for forty days

Weight of each is revealed through experience; without the doubt of going above or below

On testing it shall have a yellow shade; the likes of it are not found on this plain of existence.

R2—Recipe-poem by Ḵajeh Mir ‘Ali Tabrizi (d. 1416) found in the treatise *Medād al-Ḵoṭuṭ* by Mir ‘Ali Heravi (16th C):

دوده شش کن هریکی صمغ و نبات و صبر نه زعفران یک جزو کن زنگار دو، نیکو بسای

جمله را حل کن به آب مورد و مقداری گلاب گر همی خواهی مطوّس ساختن این است رای

Use six parts soot and nine parts of gum, rock sugar, and aloe. One part saffron, two parts Verdigris, and beat well.

Dissolve the sentence above in extract of myrtle and some rose water. If you want to make Peacock (Moṭavvas) ink, this is the way.

R3—Recipe from the treatise *Resāle dar Bayān-e, Kāgạḏ, Morakkab, va Hall-e Alvān* by anonymous (15th C):

اگر می خواهند مداد سیاه روان براق پیدا سازند و زیادت از عادت بدان مهم پردازند، خود ساختن اولی می نماید. و متقدمان در باب آداب مرکب نسخه های منتخب و مجرب مرتب کرده اند، و آنچه از همه بهتر و آسانتر بود این است. باید که قدری روغن کتان خالص بستاند و از پنبه نو، فتیله سطبر بتابد و اندک نرم کند و در چراغ نهد و روغن برسازد و در کوره ای نهد که باد نباشد، و روشن کند و سبویی آب نا رسیده را پاره ای از طرف سر بشکند و بر سر آن چراغ بیامیزد تا آنگاه که دوده جمع شود آن دوده را از سفال به پر مرغ فراهم آرد و در میان کاغذ کند و محکم بپیچد و در میان خمیر گیرد و در تنور برد تا نیک پخته شود. پس چون از آن چربی که اصل دوده مستودع است، به شدت حرارت بسوزد از تنور بیرون آرد و مقدار ده درم بر کشد و بنهد و صمغ عربی سفید پاک که اگر یک حبّه در دهن گیرد در دم آب شود و هیچ جرم نماند مقدار بیست درم بستاند و سه شبانروز در آب جوشیده کند که سرد شده باشد تا نیک حل گردد، آنگاه به کرباس محکم بپالاید. پس دوده را در هاون کند و بدان آب صمغ خمیر کند و بسیار بکوبد تا هر دو مضمحل شوند. بعد از آن مازوی رسیده بی سوراخ را نرم کند، مثل جو و گندم، پانزده درم و پنج روز در آب کند و در آفتاب نهد، تا شیره وی تمام بیرون آید. آن را نیز به رکویی سطبر بپالاید، و از این آب مازو، اندک اندک در آن دوده می ریزد و صلایه می کند، تا تمام آن آب مازو را در وی صرف کند، آنگاه پنج درم زاگ ترکی را بر روی صفحه آهن یا مس نهد و بر سر آتش دارد تا آن گوگردی که در ذات زاگ ودعیت و طبیعت است، پاک بسوزد. پس نرم بساید، و اندک اندک در سیاهی می ریزد، و چند روز به انجام، بر دوام صلایه می کند تا اجزاء مجموع بهم ضمّ شود. پس قدری آب حنا و آب برگ مورد و آب وسمه و قدری گلاب و عرق نسترن و با آب زعفران و صبر سوده و نمک فسانی و اندکی مروارید و مرجان سوده و مشک و عبیر اشهب و زر و نقره و مس و برنج حل و شنجرف و لاجورد با آن سیاهی بیامیزد که اینها هر یک خاصیتی دیگر دارند و فایدتی دیگر. و این ترکیب عجیب هرگز به سبب آب و هوا تبدیل و تغییر نپذیرد و سالهای بسیار و قرن های بی شمار بر صفحات روزگار ناهموار، پایدار و برقرار بماند. و این را مداد طاوسی میخوانند و دیر غلیظ می شود. و اگر به سبب مرور کرور شهور و اعواب و حرارت و یبوست هوای صبح و شام، اندک غلظتی در وی مشاهده کند و ناروان رود چاره آنست که اندک کفک دریا و صمغ و سندروس سوده در دوات کند تا غلیظ تر شود. پس اندکی گلاب در وی ریزد تا باز به حال اصل آید.

'In order to create a high quality fluent shiny ink, the most important thing is to make it yourself and the masters and scholars have gathered a variety of recipes and the best and easiest method is as follows: Some pure linseed oil (*Roḡan-e Katān*) is required along with a thick wick (*fitile*) made from moistened fine cotton. The oil and the wick are inserted in a lamp and turned on, put in an earthenware container so wind does not affect it, with a broken piece of clay tile on top to capture the lamp black. Afterwards, the soot is scraped off with the aid of feathers, wrapped around tightly with paper, secured in a paste (dough), and baked in a furnace. With the high heat of the furnace, the residue oil of the soot will be burnt away.

Lamp black is then taken out of the oven and 10 *Derams* (150 g) of this soot along with 20 *Derams* (300 g) of pure white gum Arabic (which if a piece of it is put on tongue, it immediately gets dissolved without leaving any traces of material), is placed in previously boiled water (now cooled) and allowed to dissolve over three days and filtered through fine silk afterwards. In the following stage, the lamp black and the liquid gum is placed in a mortar and set to bray until the particles are distributed evenly and a paste is made. Meanwhile 15 *Derams* (225 g) of ripe gallnuts without holes are ground and powdered like wheat, placed in water for 5 days and put under sunlight in order to soften the nuts and remove the extract. Afterwards the liquid is filtered and added to the lamp black in small portions and brayed until the extract is completely absorbed.

In the next step, 5 *Derams* (75 g) of Turkish vitriol (*Zāj-e Torki*) is put on an iron or copper sheet and placed on a flame until the sulphur that resides naturally inside it is burnt away.[2] The remains are then brayed slowly and added to the lamp black in small portions and set to bray for few days. Henna extract, myrtle extract, woad extract, a bit of rose water, dog rose extract (*'Araq-e nastaran*), saffron water, ground aloe, salt, a small amount of ground coral and pearl, musk, liquid amber (*'Abir-e Ašhab*), gold, silver, copper, brass, cinnabar (*Šanjarf*), and lapis lazuli (*Lājvard*) are added to the lamp black for each has its own properties and virtues.

This type of ink would never be affected by the changes of weather and will remain on the papers of turbulent eras, for many years and centuries. They call this blend the Peacock (*Ṭavusi*) ink and it thickens at a much slower rate. Should the ink thicken and lose its fluidity due to different factors such as changes in weather, temperature and time, the solution is to add some ground coral (*Kafak-e daryā*), gum, and sandarac (*Sandarus*) to make the ink thicker, and finally adding some rose water to bring it back to its original form'.

[2] The act of heating probably removed the residing water rather than sulphur, which is confirmed in an interview with Dr. Jalil Jokar (2015).

R4—Recipe from the treatise *Resāle dar Bayān-e Ḵaṭ va Morakkab va Kāġaḏ va Sāḵtan-e Ranghā* by Anonymous (15th C):

بدان که مرکب را انواع است، نوع اول و مرغوب مرکب صمغی است.

هم سنگ دوده زاج است، هم وزن هردومازو   هم وزن هرسه صمغ است وانگاه زوربازو

دوده را در ته سفالینه باید گرفت، و بهتر آن است که روغن از او بگیرند و در کاغذی کردن و آن کاغذ را در خمیر گرفتن و در تنور گرم بر سر خشت پخته نهادن، چندان که خمیر پخته شود، آنگاه روغن از او رفته باشد، پس دوده را از کاغذ بیرون آورد و صمغ عربی پاکیزه پر زور در ظرفی کند و آب بر آن ریزد آن مقدار که داند از آن آب به قوام عسل می آید، و بگذارد که چون عسل شود. پس دوده در هاون کند و قدری از آن صمغ خیس خورده که چون عسل بقوام آمده باشد بر سر دوده ریزد تا خمیر شود، و بسیار بکوبد. بعد از آن مازوج خرده را در ظرفی کند و ده چند آب در او کند و برگ حنا و برگ مورد، از هریک درمی، و نیم درم وسمه، و نیم درم افتیمون بر سر مازوج ریزد و یک شبانه روز بگذارد، و بعد از آن به آتش بجوشاند چندان که آب مازوج بر کاغذ نشو نکند. پس آنگاه بکشد و به کرباس نو بپالاید و صاف نموده آب مازوج را نگاه دارد. بعد از آن زاج قبرسی را در آب کند تا حل شود و صاف نموده به آب مازوج ملحق سازد و یک روز بگذارد. و بعد از آن صاف آب مازوج و زاج را دیگر باره صاف کند چنان که دردی در او نماند و قدری از آن در هاون بر سر دوده و صمغ ریزد. و این مقدار که سحق توان کرد سحق کند. بعد از آن پاره ای نیل سراب به اندکی صبر در هاون اندازد و سحق کند تا صد ساعت که تخمین پنج شبانه روز خواهد بود، بعد از آن تمام باشد و از این کمتر سحق ناتمام باشد.

آنگاه قدری نمک هندی با نبات مصری در هاون ریزد و همه آب زاج و مازوج را اندک اندک در هاون می ریزد و می ساید و تجربه می کند تا به حد مطوسی رسد. بعد از آن از هاون بیرون آورده، بپالاید به حریر، و نیم درم مشک تبتی و یک مثقال زعفران و ده درم گلاب حل کند که بغایت صاف شود و صاف مشک و زعفران را در مرکب ریزد و کتابت کند که بغایت خوب و روان و مطوس باشد.

Know that inks are of different types, and the highest quality is Gum ink.
The equal of soot is vitriol, the equal of both combined is gall,
the equal of all three is gum, and so the strength of muscle.

The soot must be scraped off the bottom of the earthenware container and the best soot is one free from oil by wrapping it with paper, securing it in dough then placing it on top of baked bricks in a warm oven/furnace to the point that the dough will be baked, and the residue oil will be removed, and the pure lamp black will be transferred to a mortar. Enough water is added to gum Arabic for it to take the form of honey upon dissolving. Add some of the gum Arabic solution to the lamp black until a paste is made and pound. Now take pounded gallnuts and pour water on top and add 1 *Deram* (15 g) of henna leaves, 1 *Deram* (15 g) of myrtle leaves, ½ *Deram* (7.5 g) of woad and ½ *Deram* (7.5 g) of epithymum.

Leave it overnight and boil it the next day to the point that when tested on paper, it doesn't pass through the other side and filter this solution after testing. In the following stage, mix Cyprus vitriol with water then filter and mix it will gall water and let it stand overnight. Then filter the mixture and add some to the mortar containing soot and gum Arabic and bray as much as possible. Finally, add some indigo and aloe to the mortar and bray for another 100 h or roughly 5 days. If you bray and mix less than the said amount, your ink will be unfinished.

In the next stage, add Indian salt and Egyptian sugar candy to the mortar. The rest of the gall and vitriol mixture is added as well in small portions and brayed well and tested on paper until the desired colour and consistency is reached. Once this is done, filter the ink through tent cloth and dissolve ½ Deram (7.5 g) of Tibetan musk and 1 *Meṯqāl* (4.6 4 g) of saffron in 10 *Derams* (150 g) of rosewater and add this potion to your ink and write with it for it shall be a high-quality smooth Peacock ink'.

R5—Recipe from the treatise *Ādāb-e Ḵaṭ* by ʻAbdollāh Ṣeyrafi (16th C):

هم سنگ دوده زاج است، هم سنگ هر دو مازو   صمغ است مثل هر سه، وانگه زور بازو

اما کیفیت او آنست که دوده را در کاغذ نهد و کاغذ را در خمیر سخت گیرد، و در تنور گرم نهد بر خشتی پخته. چندان که آن خمیر پخته شود. پس بیرون آرد و دوده را در هاون ریزد، و بعد از آن صمغ عربی را در ظرفی ریزد و آب بر سر آن، آن مقدار کند که چون صمغ بگدازد به قوت عسل باشد. و چون بگدازد بپالاید و در هاون ریزد، و بر سر دوده بسیار بمالد و بکوبد. بعد از آن مازو را بشکند مقدار نخود، ده چندان آب بر وی ریزد. بعد از آن برگ حنا و برگ مورد از هر یک، یک درم و نیمه نیم درم افتیمون بر سر مازو ریزد، و یک شبانه روز بگدازد. بعد از آن بر سر آتش نهد و می جوشاند و تجربه می کند چندان که آب مازو بر کاغذ نشو نکند. پس فرو گیرد به کرباس نو بپالاید. بعد از آن زاج را صافی کند و بر سر آب مازو ریزد و یک روز بگذارد. بعد از آن صافی کرده پاره پاره در هاون می ریزد و سحق می کند و تجربه می نماید، پس اندکی نیل سرآب در هاون اندازد. و باید که صد ساعت تمام سحق کند خواه در پنج شبانه روز متصل، و خواه بیشتر، که سحق تمام صد ساعت است. و چون سحق تمام شده باشد قدری نمک با نبات مصری در هاون اندازد تا بگدازد، بعد از آن بیرون آرد و به حریر بپالاید، بعد از آن اندکی مشک با گلاب حل کرده در والا اندازد و بیفشارد، و بر سر مداد ریزد و کتابت کند که مدادی به غایت خوب و مطوس باشد.

The Equal of soot is vitriol, the equal of both combined is gall, the equal of all three is gum, and then the power of muscle.

And the method is that lamp black is wrapped tightly in paper and the paper is secured inside a dough after which it is placed in a heated furnace. Once the dough is baked, it is taken out of the oven and the lamp black inside is poured into a mortar. The next step is to put gum Arabic into a bowl and add enough water for the gum to dissolve and take on the form and consistency of honey. Once the gum-water is ready and purified, it is poured on top of the lamp black and is beaten, ground and set to bray. The gallnuts are then broken into small pieces, placed in a bowl and 10 portions of water added along with henna and myrtle leaves, one *Deram* (15 g) each, and quarter of a *Deram* (3.75 g) epithymum, and it is then kept for a full day and night.

In the next stage, this mixture is boiled and tested on paper to the point that the liquid does not leak from the verso. The boiled gall water is then filtered through cloth. The vitriol is passed through a sieve and added to the liquid and kept for another day. This mixture is then filtered again, poured bit by bit into a new mortar with a small portion of indigo which is then rubbed and beaten for 100 h over five or more days if you please for the completion of braying is 100 h. Once this process is completed, a bit of salt and Egyptian sugar-candy is added and mixed till it dissolves. The content of the mortar (gall water and vitriol) is filtered through new silk. Some musk and rose water are mixed until it dissolves and is added to the mixture described above. The mixture is then poured over the lamp black and set to bray. Now write with this ink for it is a long-lasting Peacock ink'.

R6—Recipe-poem from the treatise *Resāle dar Bayān-e Tariqe-ye Sāḵtan-e Morakkab-e Alvān va Kāg̲adhā-ye Alvān by* Anonymous (16th C):

دوده شش کز هریک ازاو صمغ و نبات و صبر سرو   زعفران دو، نیل دو، زنگار دو، نیکو بسای

وانگهی آن جمله کن حل با گلاب و آب مورد   گرمطوس ساختن داری هوس این است رای

Use six parts soot, and for gum, rock sugar and cypress aloe, one part each. Use two parts saffron, two parts indigo, two parts verdigris, and beat well.

Then dissolve the sentence above in rose water and extract of myrtle. If you fancy making Peacock (Moṭavvas) ink, this is the way.

R7—Recipe from the treatise *Medād al-Ḵoṭuṭ* by Mir 'Ali Heravi (16th C):

و اگر کسی خواهد که مرکب مطوس ساخته شود، کیفیت آن این است که اول قدری روغن بذر و کتان اگر نباشد روغن کنجد خالص بستاند و از پنبه آب ندیده، فتیله ای سازد و تاب کم دهد در چراغ نهد و روشن کند و در گوشه ای که باد در او تصرف نکند سبو یا کوزه آب ندیده و یا دیگ آب ندیده بر سر آن چراغ نهند تا دوده بر آن جمع گردد،و چون دوده فراهم گردد او را از سفال با پر مرغ جمع آوری نماید، و اگر دوده خاک آلود باشد قدری آب در کاسه کند و دوده را در آن ریزد و پر مرغ را بر روی دوده بکشد تا خاک به ته کاسه نشیند و دوده بر سر آب بماند. پس با پر مرغ دوده را از روی آب بر دارند و در کیسه کرباس آب ندیده کند و در خمیر بگیرد و در تنور تافته که زیاده از حد نسوخته بر سر خشتی نهد تا خمیر پخته گردد، و آن چربی که اصل طبیعت دوده است، به شدت حرارت برطرف شود، بعد از آن از تنور بیرون آورد و صمغ عربی بی جرم که به رنگ نبات باشد در ظرف پاک، علیحده از آبی که جهت اجزای مدادجوشانیده خالصاً این مقدار بر سر صمغ ریزد که چون گداخته شود به قوام عسل باشد.

اما طریق طبخ آب مازو و اجزای دیگر چنان است که اگر دوده پانزده مثقال کا عبارت است از ده درم سنگ باشد زاج برابر آن کنند و بهترین زاج سفید است و امتحان خوبی آن این است که چون به زبان زنند، تند باشد و زبان را زبر کند و دهان را در هم کشد و چون او را به آب مازو ریزند به نقش تیره می شود. و بازو دو برابر دوده باشد و باید که مازو سبز و بی سوراخ باشد و او را نیکو صلایه کنند و صمغ عربی چهار برابر دوده باشد و برگ مورد و حنا و وسمه از هر یک پنج درم و نبات مصری و زعفران اعلا و نیل سراب و افتیمون از هر یک یک درم اولا مقدار پنج من آب صاف در دیگ پاک نموده و اگر دیگ سنگین باشد بهتر است بجوشانند تا نصف شود و دیگ را از آتش فرو گیرند و آنچه صمغ و زاج را احتیاج است از آن آب بر سر آنها ریزند بعد از آن اجزای دیگر را سوای صمغ و زاج و زعفران و نبات در ظرف پاک کرده، تتمّه آب جوشیده بر سر آنها ریزند و سه روز در آفتاب گذارند، پس در دیگ بریزند و بجوشانند تا چندان که او را بر کاغذ نویسند از طرف دیگر نشر نکند.

اما زاج را بعد از آنکه بر روی صفحه ای از آتش گذاشته به آتش داشته تا گوگردی که در اصل او باقی است، پاک سوخته باشد و او را به گلاب یا به همان آب جوشیده داخل کنند و بر سر آب مازو ریزند تا صاف شود، چنانکه هیچ درد به آب مازو مخلوط نباشد، بکار برند. و اگر مازو و آب ای اجزای دیگر مدتی بمانند بهتر باشد چنانکه پرده بر روی آن آب طبخ شده بسته می شود باد که او را بردارند مرتبه مرتبه، تا وقتی که دیگر بسته نشود. بعد از آن داخل دوده و صمغ نمایند بهتر باشد. و چون آب مازو بدین طریق طبخ و محافظت از او نموده باشند نیکوتر بود و مدادش در نهایت لطافت باشد.

بایست دانست که آب مازو و ای اجزای دیگر بعد از طبخ یافتن می باید که پنج برابر اجزای آب خالص بی درد بیرون آید و از بیست و پنج درم مازو و نیم من به وزن تبریز آب جوشیده بی درد خالص بگیرند. و مکرر امتحان شده که هر جزوی از اجزا بعد از طبخ هم وزن خود رطوبت دارد. مثلا اگر دوده ده مثقال بود بعد از جوشیدن بیست مثقال می شود و اجزای دیگر بر این قیاس، پس دوده را در هاون ریزند و با صمغ مخلوط کرده و سه روز محکم بکوبند. و هیچ شک نیست که هر چه بیشتر کوبیده و سحق و صلایه شود بهتر است.

بعد از آنکه بسیار کوبیده باشند از آب مازو اندک اندک در او ریخته، سحق کنند و در محل سحق قدری صبر و نیل سراب داخل نموده و نبات مصری را هم با نیم درم مشک در آب جوشیده با گلاب حل و صاف کرده در هاون ریزند و سحق کنند تا این هم تمام شود.

بعد از آن مداد را از هاون بیرون آورد به حریر بپالاید و در ظرف چینی یا زجاجی کرده کتابت کند بسیار براق و روان و مطوس، یعنی به رنگ پر طاوس نماید و مدادی باشد که به هر چه در آب افتد و تر شود هرگز از کاغذ نرود.

این مدادی است که خواجه جمال الدین یاقوت علیه رحمه الله الملک الودود ساخته اند. منقول است که به همین مداد سی بیت به یک مد کتابت فرمودند.

و قبله المتقدمین خواجه میر علی تبریزی البسه الله بحل النور واضع خط نستعلیق در خصوص ساختن مرکب فرموده است که در میان خطاطان و خوشنویسان تا حال همچون مدادی پیدا نگشته و در قدر و قیمت مانند طلای احمر است.

'And if one wants to make Peacock ink, the method is to firstly gather some linseed oil or pure sesame oil (if linseed is not available). A dry cotton wick (*fitile*) which has not been twisted too much is placed inside a lamp and turned on. The lamp is then placed in an area free of wind, and an earthenware container or pot is placed over the lamp to capture the soot. Once enough soot is created, it is scraped off with a feather, and if the soot has dust, place some water in a bowl and place the soot above the water and tease it with the feather so that the dust will start to deposit at the bottom of the bowl while the soot sits on the surface of water. Then, remove the soot with the feather and place and wrap it in a clean dry canvas and secured in dough, put the dough on a stick and place in a furnace to slowly bake it. The high heat of the furnace will burn away the residue oil of the soot. Then remove from the furnace. Place pure gum Arabic which looks like rock sugar in a clean container and add enough boiled water to it so that when the gum dissolves, the liquid takes the form of honey.

The preparation of the gall water and other constituents are in this manner that if there is 15 *Meṯqāls* (69.6 g) of soot, vitriol should be equivalent to that and the best vitriol is white vitriol. The best way to test it is to place a piece of it on your tongue, it will be hot, and makes your tongue dry and abrasive and if a piece of it is added to gall water, it will turn it dark. The gallnuts should be green and without any holes and it should be finely ground and should be twice the amount of soot (139.2 g). The gum Arabic should be four times the amount of soot (278.4 g) and for myrtle, henna, and woad leaves, 5 *Derams* (75 g) of each should be added, while for Egyptian sugar candy, high quality saffron, indigo and epithymum, 1 *Deram* (15 g) of each is needed.

Pour 5 *mans* (15 kg) of water into a clean pot and the best is a stone pot; boil until half the water is evaporated. Then take enough water for gum and vitriol and pour the water over them. Place all ingredients except gum, vitriol, saffron and sugar candy, into a single clean container. Pour the rest of the boiled water into that container and leave under sunlight for 3 days. Then boil the content and test on paper to the point when a drop is placed, it will not leak through the other side. The vitriol must be put on a hot plate placed on a fire so that the sulphur, which resides inside is all burnt out. Then it is added to rose water or the boiled water, and then poured into the gall water and filtered so that no solid particles are left. And if the extracts of gall nuts and other materials are allowed to rest it shall be better, since a (froth) layer will appear above the extracts which needs to be removed in turns. And should you prepare and protect the extracts in such way, it will result in a very smooth and high-quality ink.

It should be known that the weight of all the prepared extracts should be five times that of the weight of the pure water used, and of 25 *Derams* of gall in half *Mann* Tabriz[3] weight of water, repeated experiments show that after the extraction process, all ingredients would double their weight due to water. For example, if there are 10 *Meṯqāls* of lamp black (soot), after boiling it would turn to 20 *Meṯqāls* and so on.

The lamp black is then placed in a mortar; the gum is added to it and set to bray for at least 3 days. And let there be no doubt that the more you bray, the better the ink. In the next stage, the gall water is added bit by bit. Meanwhile, some aloe (*Ṣabr*), indigo (*Nil-e Sarāb*), Egyptian rock sugar and ½ *Derams* (7.5 g) of musk is boiled in water; then some rose water is added, and the content is filtered and added to the mortar and set to bray. In the last stage, the ink is filtered through cloth, placed in a China or glass container, and used. It shall be shiny, fluid and *Moṭavvas*, which means it resembles the colours of peacock feathers. And it's a

[3]The amount may be different because they mention the Tabrizi unit of measurement.

kind of ink that no matter how many times the document falls in water and gets wet, the ink shall not fade away.

It is said that Ḵāje Jamāl al-Ddin Yāqut (d. 1298) had written 30 verses with one dip of nib in this ink. And Mir ‘Ali Tabrizi (d. 1416), the creator of the Nasta’liq script, says that amongst scribes and calligraphers, there is no better ink, and that it’s worth is that of red gold.[4]’

R8—Recipe from the treatise *Medād al-Ḵoṭuṭ* by Mir ‘Ali Heravi (16th C):

بدان که مرکب را به انواع ساخته اند لیکن به نوعی بهتر و قسمی خوشتر بر این وجه است. ده درم دوده و ده درم زاج قبرسی، پنج درم مازو و پانزده درم برگ مورد و حنا و وسمه از هر یک پنج درم، نوشادر کانی نیم درم، صمغ عربی سی و پنج درم، زعفران یک درم، نبات مصری یک درم، اولا مقدار هفت من آب صاف در دیگ سنگی کنند و بجوشانند که سه من نیم بماند پس مازو و زاج و صمغ و برگ حنا هر یک در ظرفی کنند و چندان آب در سر هر یک بریزند که مغمور گردد و مازو را پیش از آنکه در آب ریزند بشکنند چنان که هر یک پنج شش پاره شود، پس در خریطه ای از کرباس آب ندیده کرده و در خمیر گرفته و در تنور تافته نهند یا بر سر خشتی گذارند تا پخته شود. آنگاه دوده و نوشادر را در هاون کنند و صمغ حل کرده را در وی ریزند آن مقدار که مانند خمیر گردد . بسیار بکوبند. پس آب مازو و آب زاج و آب برگ حنا و مورد را صاف کرده با یکدیگر بیامیزند و در دیگ کنند و به آتش نرم بجوشانند و تجربه نمایند که چون بر کاغذ نویسند از روی دیگر نشر نکند.

و بعد از آن از دیگ بیرون آورند و اندک اندک از آن آب بر سر هاون ریزند و سحق کنند تا این تمام شود. پس آب زعفران و نبات مصری را در آب حل سازند و صاف نموده و در هاون ریزند و باز اندک اندک از آن آب جوشیده می ریزند و صلایه می کنند تا وقتی که معلوم می شود به کمال رسیده، از هاون بیرون آورند و به حریر بپالایند و در ظرف چینی یا زجاجی کنند و به کار برند. این مدادی باشد که از کاغذ نرود هر چند که به آب تر شود. و شک نیست که هر چند صلایه بیشتر کنند بهتر شود. منقول است به یک مدّ سی بیت نوشته اند و این فقیر را اتفاق افتاده که به یک مدّ هیجده بیت کتابت کردم.

‘Know that there are many types of ink, but the best and most appropriate way is this: 10 *Derams* (150 g) soot, 10 *Derams* (150 g) Cyprus vitriol, 5 *Derams* (75 g) gall, 5 *Derams* (75 g) myrtle leaves, 5 *Derams* (75 g) woad leaves, 5 *Derams* (75 g) henna leaves, ½ *Deram* (7.5 g) Sal Ammoniac, 35 *Derams* (525 g) gum Arabic, 1 *Deram* (15 g) saffron, and 1 *Deram* (15 g) Egyptian sugar candy.

Firstly, 7 *Manns* (21 kg) of water is boiled until 3.5 *Manns* (10.5 kg/L) is left. Put gall, vitriol, gum, and henna leaves, each in different containers and add enough water to them, so that they are immersed. Before pouring water on gall-nuts, break them into pieces of five to six. Now take the soot, (which was previously placed and wrapped in cloth, secured in dough, and baked in a furnace), and place it in a mortar along with the sal ammoniac. Enough of the dissolved gum is added to the mortar to create a paste and it is then beaten well. Now the gall water, henna water, vitriol water, and woad water are each filtered, added together in a pot and boiled on a low flame and tested to the point that when placed on paper, it does pass through the verso.

In the next stage the contents are taken out of the pot and poured into the mortar bit-by-bit and brayed until all the elements are mixed well. Now the saffron water and Egyptian sugar candy are dissolved in boiled water, filtered, and added

[4] This means that Peacock ink had been used at least since the thirteenth century.

to the mortar in portions and brayed until it is desirable. Once the ink reaches its desirable consistency, filter it through silk and keep it in a glass or China container. This is an ink that will stay on the paper no matter how many times it gets wet. And let there be no doubt that the more you bray, the better the ink. It is said that you can write 30 verses with one dip of the pen in this ink, and this humble one happened to write 18 verses.'

R9—Recipe from the treatise *Osul va Qavā'ed-e Ḵoṭuṭ-e Sette by* Fatḥollāh Sabzevāri (16th C) (Fig. 4.8):

و اما مداد و آن به انواع ساخته اند و بهترین آن آنست که در زمان خواجه جمال الدین یاقوت ساخته اند. و آن بر ان وجه است. ده درم دوده، زاج قبرسی پنج درم، مازو بیست درم، برگ مورد و حنا از هر یک پنج درم، نوشادر نیم درم، صمغ عربی سی و پنج درم، زعفران یک درم، نبات مصری یک درم.

اولا مقدار پنج من آب صاف در دیگ سنگین کنند که چرب نباشد و چندانی بجوشاند که به سه من باز آید. پس مازو و زاج و صمغ و برگ مورد و برگ حنا هر یک جدایگانه در ظرفی کنند و چندان آب در سر هر یک ریزند که مغمور گردد و مازو را پیش از آن که در خواهد کرد بشکند چنان که هر یک پنج شش پاره شود.

پس از آن کرباس آب نا رسیده خریطه دوزد و دوده را در آنجا کند و در خمیر گیرد و میان تنور تافته نهد بر سر خشتی تا پخته شود. و اگر دوده خاک آلود باشد پیش از پختن باید شست.

آنگاه دوده و نوشادر را در هاون کنند و صمغ حل کرده را در وی ریزند آن مقدار که مانند خمیر گردد . بسیار بکوبند. پس آب مازو و آب زاج و آب برگ حنا و مورد را صاف کرده با یکدیگر بیامیزند و در دیگ کنند و به آتش نرم بجوشانند و تجربه نمایند که چون بر کاغذ نویسند از روی دیگر نشر نکند. و طریق شستن وی آنست که قدری آب در کاسه کند و دوده بر روی آب ریزد و پر مرغ بر روی دوده می کشد تا خاک به تک کاسه نشیند و دوده پاک بر سر آب بماند. پس به پر مرغ دوده را از روی آب بردارد و در خریطه کرده در خمیر گیرد و بپزد. چنانکه گفته شد. آنگاه دوده و نوشادر در هاون کند و اندکی بکوبد تا با یکدیگر بیامیزد. آنگاه صمغ حل کرده در وی می ریزد که مانند خمیر گردد و بسیار بکوبد. پس آب مازو و آب زاج و آب برگ حنا صاف کند و با یکدیگر بیامیزد و در دیگ کند و به آتش نرم می جوشاند و تجربه می نماید تا چنان شود که چون به آن آب بر کاغذ نویسد نشو نکند. بعد از آن از دیگ بیرون آورد و اندک اندک بر هاون می ریزد و سحق می کنند تا آن آب تمام شود.

پس آب زعفران و نبات مصری را در آب جوشیده حل کرده د و صاف ساخته، اندک اندک در هاون ریزد، و بعد از آن همان آب جوشیده می ریزد و صلایه می کند، تا وقتی که معلوم شود به کمال رسیده است از هاون بیرون آورد به حریر بپالاید و در ظرف چینی یا زجاجی کند و به وقت حاجت به کار برد.

و این مدادی باشد که از کاغذ نرود، هر چند به آب تر شود. و شک نیست که هر چند صلایه بیشتر یابد بهتر باشد.

'There are many different types of ink but the best type is one that was made at the time of Ḵāje Jamāl al-Ddin Yāqut (d. 1298) and this is how you make it: 10 *Derams* (150 g) soot, 5 *Derams* (75 g) Cyprus vitriol, 20 *Derams* (300 g) gall, 5 *Derams* (75 g) myrtle leaves, 5 *Derams* (75 g) henna leaves, ½ *Deram* (7.5 g) Sal Ammoniac, 35 *Derams* (525 g) gum Arabic, 1 *Deram* (15 g) saffron, and 1 *Deram* (15 g) Egyptian sugar candy. 5 *Manns* (15 kg/L) of water is boiled in a non-greasy stone pot until 3 *Manns* (9 kg/L) are left. Put gall, vitriol, gum, and henna and myrtle leaves, each in different containers and add enough water to them, so that they are immersed. Before pouring water on gallnuts, break them into pieces of five to six. Now take the soot and place it in a dry bag made of cloth, secure it in dough, attach it to a stick and bake in a furnace. And if the lamp black contains dust, it should be washed. And the method of washing is to pour some water into a bowl and pour the lampblack on top of the water. Rub the lamp black gently with a feather until the dust settles at the bottom and the pure soot is left on the surface of

Fig. 4.8 Peacock ink recipe R8 (left) and guidelines for preparing reed pen (right) from the treatise Osul va Qavāed-e Ḵoṭuṭ-e Sette by Fatḥollāh Sabzevāri (16th C). Manuscript 5819 - Tehran University Central Library

water; then remove the soot with the same feather. Then proceed with placing it in a bag, securing in dough and baking.

Next place the pure lamp black into a mortar along with the sal ammoniac and beat a little until mixed well. Enough of the dissolved gum is added to the mortar to create a paste and it is then beaten well.

Now the gall water, vitriol water, and henna water are each filtered, added together in a pot and boiled on a low flame and tested to the point that when placed on paper, it will not pass through the reverse side. In the next stage the contents are taken out of the pot and poured into the mortar bit-by-bit and brayed until all elements are mixed well.

The saffron water and Egyptian sugar candy are then dissolved in boiled water, filtered, and added to the mortar in portions and brayed until it is desirable. Once the ink reaches its desirable consistency, filter it through silk cloth and keep it in a China or glass container. This is an ink that will stay on the paper no matter how many times it gets wet. And let there be no doubt that the more you bray, the better the ink'.

R10—Recipe from the treatise *Savād al-Ḵaṭ* by by Majnun Rafiqi Heravi (16th C):

اما اگر خواهی سیاهی مطوس باشد طریقه او آنست که اول دوده را بگیری و در سفال آب نا رسیده، روغن او را بستانی. و طریق دیگر روغن گرفتن او آن است که دوده را در کاغذ کنی و آن کاغذ را در خمیر گیری و در تنور گرم بر سر خشت پخته گذاری، چندان که خمیر پخته شود. پس برون آری و در هاون زردی ریزی، بعد از آن صمغ عربی خوب که رنگ نبات باشد بی جرم و براق در ظرفی ریزی و آب در او کنی و آن مقدار بگذاری که به قوام نبات آید. پس قدری از آن صمغ حل شده در هاون بر سر دوده ریزی، و دوده را به آن صمغ خمیر کنی. بسیار بکوبی، و بعد از آن مازو را بکوبی. و مازوی خوب آن است که سبز و بی سوراخ باشد. و ده چندان آب بر او ریزی و بعد از آن برگ خار و برگ مورد و وسمه و افتیمون، از هر کدام نیم درم، این مجموع را بر سر مازو ریزی و یک شبانه روز بگذاری. و بعد از آن بر سر آتش گذاری و بجوشانی و تجربه می کنی تا چندان که آب مازو بر کاغذ نشو نکند. پس از آتش به زیر بگیری و به کرباس نو بپالایی، و آب صاف مازو را نگاه داری. بعد از آن زاک در آب اندازی تا حل شود. بعد از آن مقداری آب مازوی بی دوده بر او ریزی. بعد از آن قدری آب در هاون ریزی و صلایه کنی. پس پاره ای نیل سرآب با اندکی صبر در هاون اندازی. و می باید که صد ساعت آن مداد را صلایه کنی، و صبر کنی تا آن صلایه تمام شود خواه در پنج شبانه روز، خواه بیشتر، که صلایه تمام آنست که از صد ساعت کمتر نباشد.

چون صلایه تمام شود قدری نمک هندی و نبات مصری در هاون اندازی و صلایه می کنی و تجربه می کنی تا به حد مطوس رسد. یعنی رنگ طاووس نماید. بعد آن از هاون بیرون آری و به حریر بپالایی و نیم درم مشک با مثقالی زعفران در ده درم گلاب ضمّ کنی چنانکه جرم آنها همه حل شود. پس از آن بر وال ریزی و بیفشاری و صافی او را در سیاهی ریزی و کتابت کنی.

'If you would want your ink to be *Moṭavvas*, the method is to firstly gather soot (lamp black) and remove its oil in a dry earthenware container. And another method for removing the oil is to place the soot in paper, then wrap the paper and secure in a dough which is attached to a stick and is placed in a furnace until baked. Then remove the soot and place it in a yellow mortar. Next good quality, pure, lustrous and sugar candy coloured gum Arabic is placed in a bowl and water is added to it and left until the gum takes the strength of sugar candy.[5] Then take some of this liquid gum, add to the lamp black, and create a paste with it and beat well.

In the next step, the gall should be pounded, and the best gall is one that is green and has no holes. Add 10 times water to the gall followed by camel thorn, myrtle leaves, epithymum, and woad leaves, ½ *Derams* (7.5 g) each, and it is kept overnight. In the next stage, this mixture is boiled and tested on paper to the point that the liquid will not pass through the paper and leak from the other side at which point it is filtered through a new tent cloth. The vitriol is then dissolved (mixed) in water and some gall water is added to it. Some water is then added to the mortar and is set to bray. In the next step some indigo and some aloes are added to the mortar, set to bray and the ink is brayed and pounded for at least 5 days and 5 nights; for the braying is one that is not lower than 100 h.

Once all this is done, a small amount of Indian salt, as well as Egyptian rock sugar is added to the mortar and set to bray and tested on paper until it becomes *Moṭavvas,* meaning it resembles the colour of peacock. The contents of the mortar are filtered and ½ *Derams* (7.5 g) of musk and a *Meṯqāl* (4.64 g) of saffron are dissolved in 10 *Derams* (150 g) of rose water, filtered, added to the ink and is ready to write with'.

[5] Possibly a mistake as all other recipes mentions the form/strength of honey. Sugar candy is hard.

R11—Recipe from the treatise *Savād al-Ḵaṭ* by Majnun Rafiqi Heravi (16th C):

بستان مازوی سبز چهار سیر، و صمغ عربی چهار سیر، و دوده چراغ یک سیر، و زاک ترکی دو سیر، و یک سیر آب در چهار سیر مازو کن، و هشت سیر آب در چهار سیر صمغ، و سه شبانه روز بگذار و هر روز یکدو نوبت صمغ را بشوران. بعد از آن مازو را با یک من و نیم آب در دیگ بکن و به آتش نرم بجوشان تا به نیمه باز آید. پس در وی صمغ صاف کن بهم برآر، و زاک در هاون نرم بسای و به کرباس محکم میریز، و در آن آب انداز و بهم برآر. بعد از آن دوده را در هاون کن و قطره قطره در وی می چکان و نرم صلایه می کن تا تمام کشته شود. باز آب مازو، قطره ای چند در وی ریز و صلایه می کن تا سه نوبت. بعد از آن مازو و زاک و صمغ را قطره قطره در آن میان می ریز و صلایه می کن تا آن آبها را بخورد. بهد از آن در شیشه کن و نیم سیر نبات مصری و نیم مثقال زعفران و قدری صبر با یکدیگر بسای و با گلاب حل کرده در شیشه پاک ریز و کتابت کن.

Acquire 4 *Sirs* (300 g) of green gall, 4 *Sirs* (300 g) of gum Arabic, 1 *Sir* (75 g) of lamp black and 2 *Sirs* (150 g) of Turkish vitriol. Add 1 *Sir* (75 ml) of water to 4 *Sirs* (300 g) of gallnuts and another 8 *Sirs* (600 g) of water to 4 *Sirs* (300 g) of gum Arabic and keep it for 3 days and 3 nights. Remember to filter and clean the gum once or twice a day. Then take the gall and place it in a pot, along with 1.5 *Manns* (4.5 kg) of water and boil it on a steady fire until the water evaporates to half its amount. The next step is to add the filtered gum and bray the vitriol in a mortar and filter it through canvas/tent cloth.

Now add the lamp black to a mortar and add drops of gall water to it and bray it slowly. Do this step 3 times. Add the gall, vitriol and gum water through drops and bray till no liquid is left. Pour this mixture in a glass container; meanwhile bray ½ *Sirs* (37.5 g) of Egyptian rock sugar along with ½ *Meṯqāl* (2.32 g) saffron and some aloe. Once all of this is brayed, dissolve them in rose water, add to the container and start writing'.

R12—Recipe from the treatise *Ḵaṭ va Morakkab* by Ḥossein'Aqili Rostamdāri (16th C):

اول دوده بگیرند، و روش دوده گرفتن آن است که چراغ را در زیر ظرف سفالین بگذارند و چنان کنند که شعله چراغ به سقف آن چراغ نرسد. و اگر چنین نکند آتش در دوده می افتد و ضایع می شود. و طرف ظرف سفال را سوراخ کنند تا چراغ فرو ننشیند، بعد از آن دوده را در کاغذ درشت پیچیده، اطراف آن را دو انگشت خمیر بگیرند و میان آتش نهند تا آنکه روغن خود را بگدازد. و بعد از آنکه خمیر پخته شد دوده از آنجا بیرون آورند و آن دوده را با صمغ به مقداری که تعیین خواهد شد در هاون گذارند و تا مدت صد ساعت تخمیناً صلایه کنند و هر چند بیشتر کنند بهتر باشد. و هر گاه بهم چسبیده و خشک شود آب حنا و وسمه و مورد را در هم جوشانند و قطرات چند از آن بر آن صمغ و دوده می ریخته باشند تا نرم شود، و بر این قیاس عمل کنند تا آنکه عمل صلایه تمام شود. و بعد از آن زاج سیاه ترکی را تا سه روز در آب نهند و به چوب بر هم زنند هر روزی سه بار یا بیشتر، بعد از آنکه مازو را در هم کوفته به قدر مرکب آب بریزند و در دیگ ریخته بجوشانند تا ثلثان برود و ثلث بماند. و هر گاه که از آن آب بردارند و بر کاغذ نویسند منتشر نشود. این نیز علامت اتمام طبخ است. بعد از آنکه از دیگ پایین آورند و زاج را با مازو بهم آمیزند و اگر خواهند مرکب به آب ضایع نشود آب افتیمون و آب پوست گردکان در میان آن ریزند. چه بعد از این مرکب به شستن ضایع و باطل نشود، لیکن خطاطان به واسطه آن که عمل مذکور مرکب را غلیض می کند و به واسطه آن خط ناصاف می شود آب افتیمون و مثل آن در مرکب نریزند. بعد از آن آب زاج مخلوط با مازو را در میان صمغ و دوده صلایه کرده، ریخته در میان کیسه ای که از دو پاره سخت بهم گرفته و بخیه ها بر او زده باشند صاف کنند مرکب اعلا حاصل می شود. و این قسم مرکب را طاوسی گویند چه الوان نماید.

'First create lamp black and the method is to place a lit lamp underneath an earthen container and make sure the flame is not high enough to reach the top for that can burn the soot and destroy your ink. Make a hole in the earthenware container to allow air circulation and prevent the flame from going out. In the next step gather your lamp black, place on paper and wrap it and secure it in dough with thickness of 2 fingers. Place the dough in a fire and allow the residue oil to be burnt away resulting in pure lamp black. The prepared lamp black is removed and placed in a mortar along with some gum and is set to bray for an average of 100 h, for the more you bray, the better the quality.

If the lamp black and gum becomes too dry or they start sticking to each other, henna water, myrtle water and indigo water are boiled together and drops of this potion is added to the mortar to soften the ink. This process is done over and over till the mixing is complete. In the next stage, some black Turkish vitriol is placed in water for 3 days and stirred 3–4 times a day with a wooden stick, while the gallnuts are beaten, and water is poured on it and set to boil until the water is evaporated to a third of its original amount. The liquid is then tested on paper with a reed pen to the point that it will not bleed and pass through the sheet, this is a sign of completion of evaporation. Now the vitriol and gall are mixed together and if you want to make sure that the liquid will be of good quality, add some extract of epithymum and extract of green walnut husks. This will ensure that the ink will not fade or rub away upon contact with water.

Many calligraphers however have advised against using the likes of epithymum extract as it can make the ink too thick and as a result, makes the script uneven. The mixture of gall nut extract and vitriol water is added to the lamp black and gum and is set to bray and poured into a bag that is made from 2 separate sheets of cloth which are sewn together allowing slow filtration. The result is an ink of utmost quality, and they call it *Ṭavusi* (Peacock) Ink because it looks colourful'.

R13—Recipe from the treatise *Ādāb al-Mašq* by Bābā Šāh Iṣfahāni (16th C):

چون ده مثقال دوده چربی گرفته به دست آورد باید که چهل مثقال صمغ را در آب کند تا چون عسل شهد شود، و آن دوده را در کاسه کند و چهار دانگ صمغ را هم در کاسه کرده آن مقدار کف بمالد که آن دوده کشته شود. و بعد از آن، آن دو دانگ صمغ را یک مثقال آب اضافه کند تا روانتر شود و آن را نیز در آن کاسه ریزد و با آن دوده بیامیزد. چنان که آن را در شیشه توان کرد، پس آن را در شیشه یا در ظرفی که مثل آن باشد کند چنان که تا نصف آن بشود و سر آن را محکم بگیرد و آن را بجنباند تا کمال آمیختگی پیدا کند و هرچند بیشتر بجنباند صلایه بیشتر خواهد یافت. واقل مرتبه ده روز باید جنبانید. و بعد از آن چهل مثقال زاج سیاه را در آب کند تا گداخته شود و صاف کند و بعد از آن هشتاد مثقال مازو را خرد کرده با نیم مثقال زعفران و دو دانگ صبر و دو مثقال افتیمون و ده مثقال پوست گرد کان سبز و دو مثقال نیل را در نیم من آب بجوشاند تا به نصف آید و آن را صاف کند و از آن آب زاج اندک اندک در او ریزد و قلم بر آنجا زده، نویسد و تجربه کند تا رنگ آن فریسه ای شود. و این را نیز اگر اندک صمغی اضافه کند مرکبی فریسه ای خواهد بود. غرض که اعتدال او آن است که فریسه اعلا شود. و واقف باشد که زاج زیاد نباشد که مرکب را می سوزد. و بعد از آن مرکب را از شیشه بدر آورد و ازین آب زاج و مازو اندک اندک بر او ریزد و تجربه کند تا به اعتدال آید.

اما باید که صمغ و دوده که در شیشه می کند که بجنباند، چنان نباشد که قبول لت نکند، یعنی خیلی غلیظ باشد والا چندان که بجنباند صلایه نخواهد یافت. و این نوع مرکب ساختن چند فایده دارد که آن نوع که در هاون می سازند ندارد. یکی آنکه چون در اندرون شیشه است از گرد و خاک محفوظ است. دیگر آنکه همه آن مرکب به یک نوع صلایه می یابد برخلاف هاون که بعضی صلایه می یابد.

'Once you gather 10 *Meṯqāls* (46.4 g) of oil free lamp black, add 40 *Meṯqāls* (185.6 g) of gum to water until a honey like syrup is made. Put the soot into a bowl and add four sixths (4/6) of the gum water to it and mix and rub well until the soot is no longer in powder form. Add a *Meṯqāl* (4.64 g) of water to the remaining gum (2/6) to make it more fluid, then and add to the lamp black and bray once again, before pouring the ink into a bottle. The bottle should be big enough that the soot mix will fill not more than half its volume. Now seal the top of the bottle and shake well to mix the particles evenly and the more you shake the better. Ideally, it is best to shake and mix for ten days.

The next step is to dissolve 40 *Meṯqāls* (185.6 g) of vitriol in water and filter it. Meanwhile take 80 *Meṯqāls* (371.2 g) of broken gallnuts, along with ½ *Meṯqāls* (2.32 g) of saffron, some aloe, 2 *Meṯqāls* (9.28 g) of epithymum, 10 *Meṯqāls* (46.4 g) of green walnut husks, and 2 *Meṯqāls* (9.28 g) of indigo and boil it in ½ *Mann* (1.5 kg) of water until the liquid evaporates to half its amount; at which point it is filtered and the vitriol water is added in small portions. Test this ink until you see a blue-grey shade (*Fariseyi*). Remember that if you add too much vitriol, it shall destroy the ink. Afterwards pour the ink out from the bottle and into a container and add the gall water and vitriol water in small portions and test them. Using a bottle has two benefits that a mortar doesn't: the bottle has a lid which keeps dirt away from the ink; and when you shake the bottle, you are mixing all the ingredients in a continuous uniform manner whilst this is not the case in a mortar'.

R14—Recipe from the treatise *Majmu' al-Ṣanāye'* by Anonymous (16th C):

بیارند دوده نفت یا دودهٔ کتان ده مثقال، و صمغ عربی خالص چهار مثقال، مرقشیشای سوخته پنج مثقال، و صبر سقوطری دو مثقال، جمله را بسایند و نمک و روغن بگیرند که اگر روغن در او باشد مداد را خراب کند و نیک نیاید و در نوشتن از هم برود و بیاض میان حروف پیدا شود و خوش آینده نباشد. طریقش آن است که کیسه ای از کاغذ بدوزند پهن، و دوده را در میان کیسه مذکور کنند و کیسه را در میان خمیر گندم گیرند. و خمیر باید که بسیار نرم نباشد و سخت نیز نباشد بلکه موازنه باید که کیسه را در خمیر کند غلیظ باشد یا آتش به کاغذ زنند که دوده ناچیز شود. کیمه مذکور را در دکان خباز در تنور بندند یا در شیب آنشگاه یا در شیب خاکستر گرم دفن کنند تا آن خمیر پخته شود بلکه خمیر نیم سوخته شود، بعد از آن بردارند و بگذارید تا سرد شود، دیگرمرقشیشا در آتش تیز اندازند تا سوخته گردد بردارند و سرد کننده و دوده را در هاون کنند و نیم روز سحق کند و صمغ را در آب تر کند و صافی سازد. و به هژده درم صمغ صد درم آب باید کرد و بعد چندان آب کنند که دوده در او سرشته شود، بکوبند تا نیم روز، و مرقشیشای سوخته در آن افکنند و نیم روز دیگر بکوبند و اندک اندک آب صبر محلول در آن ریزند تا تمام آب صمغ بکار رود، چون صمغ بکار برده باشند نیم روز دیگر اجزاها را بکوبند، پس بستانند برگ مورد با برگ حنا صد درم و در آب کنند تا چون نرم شود بعد از آن در پاتیله کنند و بجوشانند یک ساعت تا با هم آید، پس فرو گیرند و بگذارند تا خوب سرد شود، بعده بپالایند و صاف او را در ظرف دیگر کنند و اندک اندک در هاون بر سر آن اجزای مسحوق می ریزند تا آب مورد تمام شود. اگر رقیق و روان شده باشد بردارند و در ظرف آبگینه کنند و یا در مینه، و بکار برد. و اگر رقیق و روان نباشد قدره ذره ای گلاب خالص اندک اندک بر سر آن کنند و خق کنند و به قلم تجربه کنند تا آنگاه که نیک روان شود، چون معلوم شد که نیک روان است در ظرف آبگینه یا مینه کند، محکم ببندد چنانچه اگر رقیق باشد، غلیظ گردد و از نو صلایه باید کرد این نوع مداد در غایت خوبی و نیکویی باشد، و اگر عیاذ بالله کتاب او در آب افتد اولی باید تا سیاه نشود، از جای خود نرود و برکنده نشود.

'Gather 10 *Meṯqāls* (46.4 g) of soot made of linseed or petroleum oil, 4 *Meṯqāls* of gum Arabic (18.56 g), 5 *Meṯqāls* (23.2 g) of burnt marcasite (*Marqešišā* / iron sulfide) and 2 *Meṯqāls* (9.28 g) of Aloe Vera. Bray each ingredient of the sentence above and make sure the residue oil within the soot is removed. The method for removing oil is to place the lamp black in a bag made from paper and place it inside dough made of wheat. The dough needs to be of balanced consistency (not too soft or hard). Place the said dough in a furnace or bury it under hot ashes until the dough is half burnt; then let it cool down. Meanwhile the marcasite is placed on a high flame, burnt and then allowed to cool down. The lamp black is placed in a mortar and is beaten and brayed for half a day. Water is added to the gum Arabic while the filter is prepared.

And remember that for each 18 *Derams* (270 g) of gum Arabic, 100 *Derams* (1.5 kg) of water should be added to to the point that the soot is turned into a paste; beat for half a day. Add the marcasite and beat for another half day whilst adding aloe water in drops until it is all used up; then beat for another half day.

Gather 100 *Derams* (1.5 kg) of myrtle leaves and 100 *Derams* (1.5 kg) of henna leaves and place in water to soften them. Then boil it for an hour and let it cool. Filter it and add to the mortar in portions until no more is left. If the ink is fluid and smooth, add to a glass container and use. If not, add a bit of pure rose water in drops and set to bray again until the right smoothness and fluidity is reached. It shall be an ink of the utmost quality. Should your book fall into water, the pages will not darken, and the ink shall not fade away or break away from the substrate'.

R15—Resāle dar Bayān-e Tariqe-ye Sāḵtan-e Morakkab-e Alvān va Kağaḏhā-ye Alvān by Anynymous (16th C):

منشی باید که فوایض و محاسن و معایب قلم داند آنگاه ترتیب و ترکیب مرکب و مداد سیاه روان لایق پیدا کند خود ساختن اولی می نماید و متقدمان در باب آداب مرکب سخنها مرتب و مرگب کرده اند اما آنچه همه گفته اند آن است که باید اول قدری روغن کتاني خالص بستاند و از پنبه نو فتیله ای سطبر بتابد و اندک نم کند و در چراغ نهد و پر روغن سازد و در گوشه ای نهد که باد در او تصرف نکند، و سبوی آب نادیده را پاره ای از سر آن بشکند و بر سر آن چراغ نهد تا آنگاه که دوده جمع شود، آنگاه آن دوده را به پر مرغ از سفال فراهم آورد و در میان کاغذنهد و محکم بپیچد و در میان خمیر گیرد و تنوری که پر هیمه یافته باشد، در زیر خاکستر کند تا آنگاه که مثل نان پخته شود. و در اصل طبیعت دوده ودیعت است به شدت حرارت بسوزد، پس از تنور بیرون آورد، مقدار ده درهم برکشد و بنهد، وصمغ عربی سفید پاک که اگر از وی حبه ای را گند، و در دهان گیرد و هم در آب اندازد از وی هیچ جرم باقی نماند مقدار هشت درهم بستاند و سه شبانه روز در آب جوشانیده، سرد کرده، افکند تا نیک حل شود و به کرباس محکم بیالاید. پس دوده را به آب در هاون روهین حل کند و خمیر کند، پس بسیار بکوبد تا هر دو بهم آمیخته شود. بعد از آن پانزده درهم مازوی بی سوراخ را خرده کند که همچون جو کند و در آب کند و پنج شبانه روز در آفتاب نهد تا شیره آن پاک بیرون آید و آن را به رکوی بیالاید و آب آن را در دیگ سنگین کند و آتش را نرم کند تا وقتی که چنان بقوام آید که اگر بر کاغذ نویسند نشر نکند پس پنج درهم زاج ترکی بر روی صفحه ای از آهن نهد بر آتش تا گوگردی که در اصل طبیعي او ودیعت است پاک بسوزد. پس آنگاه در گلاب حل کند اندک اندک در سیاهی بریزد و به آن آب ماز و صلایه کند، چنانچه بقرار رسد و اجزاء مجموع بهم مختلط شود، و آنگاه قدری آب حنا و آب وسمه و آب تخم مورد و گلاب و عرق کاسنی و زعفران و اندکی کف دریا و صبر سوده و نمک سوده در وی ریزد که اینها هریک خاصیتی دیگر دارد و نیکو صلایه کند.

و این ترکیب عجیب و غریب است و هرگز به سبب آب وهوا تغییر و تبدیل نمی یابد و سالهای بسیار و قرنهای بیشمار ادر صفحات روزگار پایدار می ماند. و این را مداد کامل طاووسی می گویند، و دیر غلیظ می شود. و اگر به سبب دور و گردش شهور و ایام یبوست و حرارت هوا و صبح و شام غلیظی در مزاج آن مشاهده کند و ناروان شود، چاره آنست که اندک کف دریا و صمغ عربی به گلاب حل کند و در دوات چکاند تا باز به اصلی خود آید.

'A penman must know the grace, benefits, and deficiencies of his pen as well as the constituents and composition of inks and he must find an ink which is black, fluent and worthy of writing. To achieve this, it's best to create your own ink and the ancient ones (calligraphers) have much to say on the different manners of ink making, however they all agree on the following. The first thing to do is purchase pure linseed oil and make a thick wick out of fresh cotton and moisten it and place it in the lamp along with the oil. The lamp is placed in a corner free from wind and an earthenware plate is placed above the lamp to capture the soot, which is then scraped off with a feather. The scraped off soot is placed in a paper bag, secured in dough, and placed under the hot ashes of a high flamed furnace and baked like bread. Due to this heat the residual oil will be removed and the amount of 10 *Derams* (150 g) of pure lamp black is placed in a mortar.

Next, 8 *Derams* (120 g) of gum Arabic (which if a piece of it is placed on tongue or in water, it will dissolve and leave no traces of extraneous material) is placed in cold water which was previously boiled, for 3 days and 3 nights and allowed to dissolve and finally filtered through tent cloth. In the next stage, the lamp black is poured into a zinc (*Ruhin)* mortar along with the gum water and is brayed to make a paste first, then beaten well until all the elements are mixed together. Afterwards 15 *Derams* (225 g) of pure gallnuts (without holes) are brayed like barley and added to water and set under sunlight for 5 days until full extraction. The liquid is filtered and set to boil in a stone pot after which it is tested on paper to the point that it does not pass through the other side.

Meanwhile 5 *Derams* (75 g) of Turkish Vitriol is placed on a sheet of metal and put on a fire to burn out the sulphur that resides naturally within the metal sulphate. The vitriol is then mixed with rose water and added to the soot drop by drop. Gall water is added as well, and it is set to bray to the point that they all mix evenly. Now some extract of henna, woad extract, extract of myrtle seeds, rose water, chicory extract *('araqe kāsni*), saffron, ground coral, brayed aloe and brayed salt is added and set to bray for each ingredient has its own virtue. And this is a mixture of wonders, which would never be spoiled due to changes of weather or passage of time, and it will remain on the books for many generations. They call this ink "Peacock" (*ṭāvusi*) ink, which thickens slower than others. If by any chance, some thickness is witnessed, you should dissolve some gum Arabic and ground coral in rose water and add to the ink by drops until the ink takes its natural form.

R16 Poem-recipe from the treatise *Golestān-e Honar* by Qāżi Aḥmad Qomi (16th C):

همسنگ دوده زاج است همسنگ هر دو مازو   هم وزن هر سه صمغ است آنگاه زور بازو

دوده را در ته سفالینه بایدگرفت بهتر آنست که از روغن بزرک بگیرند ودر کاغذی کردن و آن کاغذ را در خمیری گرفتن و در تنور گرم بر سر خشت پخته نهادن چندانکه خمیر پخته شود آنگاه روغن ازو رفته باشد پس دوده را از کاغذ بیرون آورده صمغ عربی پاکیزه بریزد و بر ظرفی کند و آب بر آن ریزد این مقدار که دانه از آن آب بقوام عسل می آید و بگذارد که چون عسل شود پس دوده در هاون کند و قدری از آن صمغ خیس خورده چون عسل بقوام آمده باشد و بر سر دو ده ریزد تا خمیر شود و بسیار بکوبد بعد از آن مازوج خورد کرده را در ظرفی کند و دو چندان آب درو کند و برگ حنا و برگ مورد از هر یک درمی و نیم درم و سه و نیم درم افتیمون بر سر مازوج ریخته و یک شبانه روز بگذارد و بعد از آن بجوشاند چندانکه مازوج بر کاغذ نشو نکند پس آنرا بکشد و بکرباس نو بیالاید و صاف نموده آب مازوج را نگاهدارد بعد از آن زاج قبرسی را در آب حل کند تا حل شود و صاف نموده باب مازوج ملحق سازد و یکروز بگذارد و بعد از آن صاف آب مازوج و زاج را دیگر باره صاف کند چنانکه اصلادرو دردی نماند و قدری از آن در هاون بر سر دوده و صمغ ریزد و این مقدار که سحق توان کرد سحق کند بعد از آن پارویی نیل سر آب با اندکی صبر در هاون اندازد وسحق کند تا صد ساعت که تخمین پنج شبانروز خواهد بود بعد از آن تمام باشد و ازین کمتر سحق ناتمام باشد آنگاه قدری نمک هندی با نبات مصری در هاونریزدو همه آب زاج و مازوج را اندک اندک در هاون میریزد ومی ساید و تجربه می کند تا بحد مطوسی رسد بعد از آن از هاون بیرون آورده بیالایدبحریر و نیم درم مشگ تبتی ویک مثقال زعفران درده درم گلاب حل کند بغایت صاف شود و صاف مشک و زعفران را در مرکب ریزد و کتابت کند که بغایت خوب و روان و مسطوس باشد.

The equal of soot is vitriol, the equal of both combined is gall,

the equal of all three is gum, and so the strength of muscle.

The soot must be scraped off the bottom of the earthenware container and the best soot is one free from oil by wrapping it with paper, securing it in dough then placing it on top of baked bricks in a warm oven/furnace to the point that the dough will be baked, and the residue oil will be removed, and the pure lamp black will be transferred to a mortar. Enough water is added to gum Arabic for it to take the form of honey upon dissolving. Add some of the gum Arabic solution to the lamp black until a paste is made and pound. Now take pounded gallnuts and pour water on top and add 1 *Deram* (15 g) of henna leaves, 1 *Deram* (15 g) of myrtle leaves, ½ *Deram* (7.5 g) of woad and ½ *Deram* (7.5 g) of epithymum.

Leave it overnight and boil it the next day to the point that when tested on paper, it doesn't pass through the other side and filter this solution after testing. In the following stage, mix Cyprus vitriol with water then filter and mix it will gall water and let it stand overnight. Then filter the mixture and add some to the mortar containing soot and gum Arabic and bray as much as possible. Finally, add some indigo and aloe to the mortar and bray for another 100 h or roughly 5 days. If you bray and mix less than the said amount, your ink will be unfinished.

In the next stage, add Indian salt and Egyptian sugar candy to the mortar. The rest of the gall and vitriol mixture is added as well in small portions and brayed well and tested on paper until the desired colour and consistency is reached. Once this is done, filter the ink through tent cloth and dissolve ½ Deram (7.5 g) of Tibetan musk and 1 *Meṯqāl* (4.6 4 g) of saffron in 10 *Derams* (150 g) of rosewater and add this potion to your ink and write with it for it shall be a high-quality smooth Peacock ink'.

R17—Recipe from the treatise treatises *Kašf al-Ṣanāye'* by 'Ali Ḥosseini (19th C):

هم سنگ دوده زاج است　　　هم وزن هر دو مازو

هموزن هر سه سنگ است　　　آنگاه زور بازو

هرگاه خواهی سیاهی مطوس باشد کیفیت آن آن است که اول دوده را بگیری و چربی آن را بستانی. و طریق گرفتن روغن از دوده آن است که دوده را در کوزه سفال آب ندیده بریزی و در تنور بسیار گرم گذاری تا کوزه روغن آن را جذب نماید، پس بیرون آورده در هاون روی بریزی و صمغ عربی صاف خوب را در ظرفی علیحده ریخته، آب صاف دوده ودیعت است به شدت حرارت بسوزد، پس از تنور بیرون آورد، مقدار ده درهم برکشد و بنهد، وصمغ عربی سفید پاک که اگر از وی حبه ای را گند، و در دهان گیرد و هم در آب اندازد از وی هیچ جرم باقی نماند مقدار هشت درهم بستاند و سه شبانه روز در آب جوشانیده، سرد کرده، افکند تا نیک حل شود و به کرباس محکم بیالاید. پس دوده را به آب در هاون روهین حل کند و خمیر کند، پس بسیار بکوبد تا هر دو بهم آمیخته شود. بعد از آن پانزده درهم مازوی بی سوراخ را خرده کند که همچون جو کند و در آب کند و پنج شبانه روز در آفتاب نهد تا شیره آن پاک بیرون آید و آن را به رکوی بیالاید و آب آن را در دیگ سنگین کند و آتش را نرم کند تا وقتی که چنان بقوام آید که اگر بر کاغذ نویسند نشر نکند پس پنج درهم زاج ترکی بر روی صفحه ای از آهن نهد بر آتش تا گوگردی که در اصل طبیعي او ودیعت است پاک بسوزد. پس آنگاه در گلاب حل کند اندک اندک در سیاهی بریزد و به آن آب ماز و صلایه کند، چنانچه بقرار رسد و اجزاء مجموع بهم مختلط شود، و آنگاه قدری آب حنا و آب وسمه و آب تخم مورد و گلاب و عرق کاسنی و زعفران و اندکی کف دریا و صبر سوده و نمک سوده در وی ریزد که اینها هریک خاصیتی دیگر دارد و نیکو صلایه کند.

و این ترکیب عجیب و غریب است و هرگز به سبب آب وهوا تغییر و تبدیل نمی یابد و سالهای بسیار و قرنهای بیشمار ادر صفحات روزگار پایدار می ماند. و این را مداد کامل طاووسی می گویند، و دیر غلیظ می شود. و اگر به سبب دور و گردش شهور و ایام یبوست و حرارت هوا و صبح و شام غلیظی در مزاج آن مشاهده کند و ناروان شود، چاره آنست که اندک کف دریا و صمغ عربی به گلاب حل کند و در دوات چکاند تا باز به اصلی خود آید.

The equal of soot is vitriol, the equal of both is gall,

the equal of all three is gum and so the strength of muscle.

If you want your ink to be "*Moṭavvas*" (Peacocked) first you need to remove oil from your soot and the way to do that is to place the soot in an earthenware vase, place that vase in a hot furnace, at which point the vase will absorb the oil. The soot is then removed and placed in a zinc mortar. Meanwhile place clean Gum Arabic in a separate container and add just enough water so that upon dissolving, a syrup with honey like consistency is left. This mixture is then added to the lamp black and is beaten, ground and rubbed for two to three days to allow even dispersion of soot in the gum Arabic liquid. In the next stage some no-hole green gallnuts are beaten, and 10 times water is poured on top along with 1 *Deram* (15 g) of henna leaf, 1 *Deram* (15 g) of myrtle leaf, ½ *Deram* (7.5 g) of epithymum and ½ *Deram* (7.5 g) of woad.

This potion is left for 1 day and 1 night which is then boiled and tested on paper to the point that the liquid will not pass through and leak from the other side at which point it is taken off the fire and filtered through canvas/tent cloth *(Karbā̱s)*. In the next stage some Cyprus vitriol, which is amongst the finest vitriol and has a yellow colour is added to the gall water. Remember that the best method to check whether vitriol is of high quality or not is to test it on your tongue and it shall test extremely hot/harsh *(tond)* and if you put it in gall extract, it will turn it purple. A bit of this liquid is added to the soot and set to bray (beat and mix). Note that during the braying, some indigo and aloe is added to the mortar and the liquid is added dropwise.

Now you must bray it for 100 h and the process will finish in a week. After that some Indian salt and rock sugar is added to the mortar and set to bray and mixed with the others and tested. The liquid is then filtered through silk. In the following stage, ½ *Deram* (7.5 g) musk and 2 *Meṯqāls* (9.28 g) of saffron is dissolved in 10 *Derams* (150 g) of rose water and added to the ink and it is now ready to write with for it is a lustrous smooth Peacock ink'.

R18—Recipe from the treatise treatises *Kašf al-Ṣanāye'* by 'Ali Ḥosseini (19th C) (Fig. 4.9):

نوع دیگر، دوده ده درم، زاج قبرسی پنج درم، مازو بیست درم، برگ مورد و حنا از هریک پنج درم، نوشادر نیم درم، صمغ عربی سی و پنج درم، زعفران یک درم، نبات مصری یک درم. اول مقدار پنج من آب صاف در دیگ سنگی بجوشانند تا سه من بماند پس مازو کوفته و زاج و صمغ و برگ حنا و مورد هریک را در ظرفی نمایند و چندان از آب جوشیده بر سر هریک بریزند که مغمور گردد و اگر دوده خاک آلوده باشد قدری آب در کاسه نمایند و دوده را بر آن ریزند و پر مرغ بر آن دوده کشند تا خاک آن به ته کاسه نشیند و دوده بر روی آب بماند پس با پر مرغ دوده را از روی آب بردارند و در خریطه کر باس آب ندیده کرده و در خمیر گرفته در تنور تافته گذارند تا پخته گردد و روغن آن برطرف شود آنگاه دوده و نوشادر را در هاون کرده با صمغ حل نمایند و بکوبد آن مقدار که خمیر شود پس آب مازو و زاج و حنا و مورد را صاف نموده با یکدیگر بیامیزند و در دیگ کرده به آتش نرم بجوشانند و تجربه نمایند تا به حدی رسد که چون نویسند نشر نکند. بعد از آن از دیگ بیرون آورده اندک اندک در هاون ریخته، سحقق کنند. و باید که در صد ساعت سحق آن به اتمام رسد. و شک نیست که هر چه بیشتر صلایه شود بهتر خواهد بود. بعد از آن زعفران و نبات مصری در آب جوشیده و حل و صاف نموده در هاون ریزند و سحق کنند این هم تمام شود. پس از آن از هاون بیرون آورده به حریر بپالایند و در ظرف چینی یا زجاجی کرده در وقت حاجت بکار برند. و این مدادی باشد ثابت، و محو نشود هر چند به آب بشویند. و چنین مشهور است که به یک مد آن سی بیت نوشته می شود لیکن داعی به تجربه آورده به یک مد هیجده بیت نوشته می شود. و این مرکبی است که در زمان یاقوت ساخته شده. هر گاه آب زاج و مازو که از مرکب زیاد ماند در شیشه کرده نگاهدارند بجای آب و گلاب در دوات ریزند آن را به عمل آورد از غلظت بیرون آرد.

'Another method is to get 10 *Derams* (150 g) lamp black, 5 *Derams* (75 g) Cyprus vitriol, 20 *Derams* (300 g) gallnuts, 5 *Derams* (75 g) myrtle leaf, 5 *Derams* (75 g) henna leaf, ½ *Deram* (7.5 g) sal ammoniac, 35 *Derams* (525 g) gum Arabic, 1 *Deram* (15 g) saffron, 1 *Deram* (15 g) Egyptian rock sugar.

First, boil 5 *Manns* (15 kg) of water so that after evaporation 3 *Manns* (9 kg) remains. Then the brayed gall, vitriol, gum, henna leaf, and myrtle leaf are placed in separate containers to which is added enough of the boiled water until each one is immersed. If the soot contains impurities, pour some water in a bowl and place the soot gently on top of the water and slowly graze the soot with feathers so that the dirt will sink to the bottom of the bowl (while the soot floats on top).

Remove the soot with the feather and place it in a canvas and wrap it and secure it in dough then place it in a ready heated oven and set to bake. This will burn the oil residue in the soot. Then the lamp black is placed in a mortar along with the sal ammoniac and gum water is added and set to bray till a paste is created. Now the gall water, vitriol water, henna water and myrtle water are each filtered then mixed and set to boil and tested to the point that the new liquid does not pass through the paper.

In the next stage, this liquid is added to the mortar slowly and brayed for 100 h. There is no doubt that the more it is brayed and whisked, the better. The next stage is to dissolve saffron and the Egyptian rock sugar in water, set to boil, then filter

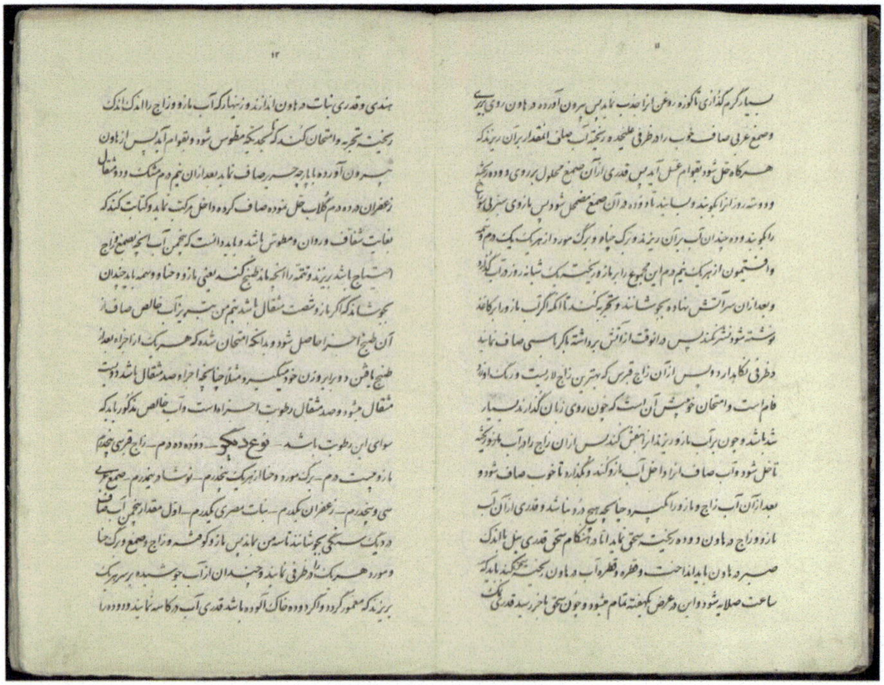

Fig. 4.9 Recipes on making Peacock Ink (R18) from the treatise Kašf al-Ṣanāye 'by' Ali Ḥosseini (19th C). Manuscript 2261 – Tehran University Central Library

and add to the mortar and set to bray. The content is then filtered through a sheet of silk (*ḥarir)* and placed in a China plate and used as necessary. This will be a stable ink that will not fade even if it encounters water. This ink is famous for being able to write 30 verses with one dip of the pen and I managed to write 18 verses. This is an ink that was created during the "*Yāqut*"[6] period. If there is extra vitriol water and gall water remaining, keep it in a glass container and you can add them to the ink instead of water or rose water if it becomes too thick; this will revert the ink back to its original fluent form'.

4.1.5 *Starch Ink*

Starch inks have unique qualities. Although most of the time the starch is burnt away and carbonised, they were used as both colorants and binders. Sometimes they were used in conjunction with other colorants and binders as seen in the following recipes. There are six recipes found in the treatise researched (Fig. 4.10; Table 4.6).

[6]Yāqūt Musta'ṣamī (d.1298).

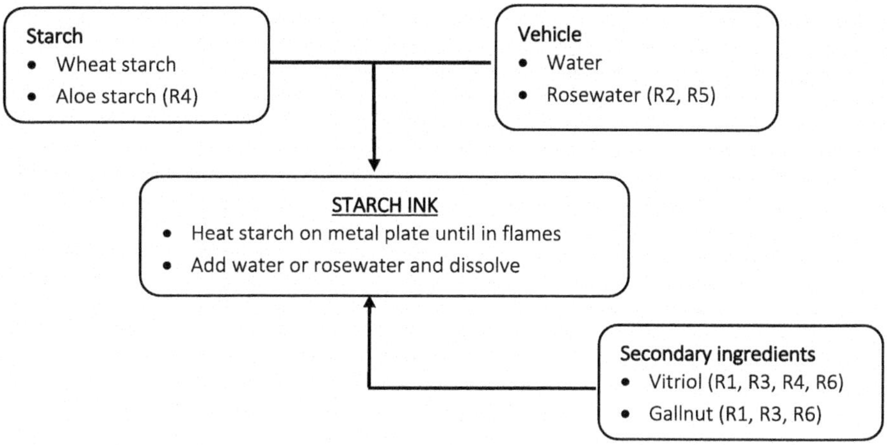

Fig. 4.10 Flowchart guide for constructing starch ink

Table 4.6 Starch Ink—ingredients and measurements

Starch ink	Primary ingredients	Parts	Amount	Secondary ingredients	Parts	Amount
R1 (12th C)	Wheat starch Gallnuts Vitriol	– – –	– – –	– – –	– – –	– – –
R2 (15th C)	Starch Water/rose water	– –	– –	– –	– –	– –
R3 (15th C)	Wheat starch Vitriol Gall Water	– – – –	– – – –	– – – –	– – – –	– – – –
R4 (16th C)	Aloe starch Vitriol Water	– – –	– – –	– –	– –	– –
R5 (16th C)	Burnt starch Rose water	– –	– –	– –	– –	– –
R6 (16th C)	Wheat starch Vitriol Gall Water	– – – –	– – – –	– – – –	– – – –	– – – –

R1—Recipe from the treatise *Bayān al- Ṣanāʿāt* by Teflisi (12th C):

حبر نشاسته: بستاند گندم و در دیگی مسین کند و آتشی نرم بزیرش برافروزد تا بریان گردد و سیاه شود و نگاهش دارد تا نسوزد. پس خردش بساید چون هبا و قدری آب مازو بر وی کند و بر آتش نهد تا بجوشد. پس بپالاید و قدری زاگ در وی کند، پس در شیشه کند و بکار برد.

Starch-Iron-Gall Ink: Acquire wheat and place in a copper pot and place on slow fire and roast it (wheat)- until it turns black and keep it until it is just about to be burnt. Then grind it well until powdered, then add to it some gallnut extract, and place on fire and set to boil. Afterwards filter, add some vitriol, pour in glass container and use.

R2—Recipe from the treatise *Resāle dar Bayān-e, Kağad̲, Morakkab, va Hall-e Alvān* by Anonymmous (15th C):

آسانتر آنست که قدری نشاسته را در تابه آهنین کند و آتشی تیز کند تا بسوزد و آتش در وی افتد. پس به آب یا گلاب حل کند، و بدان کتابت کند.

An easy method is to place some starch on an iron plate and place the plate on high heat until it is in flames. Add water or rose water and mix it until the starch is dissolved and use as necessary.

R3—Recipe from the treatise *Resāle dar Bayān-e K̲aṭ va Morakkab va Kağad̲ va Sāk̲tan-e Ranghā* by Anonymmous (15th C):

بستاند نشاستج گندم و در دیگ مسین کند ، و بر آتش نرم بریان کند تا سیاه گردد، و نگاه دارد تا نسوزد. پس بساید و قدری مازوج در آب کند تا نرم شود و بگذارد تا صاف شود، در میان نشاستج ریزد و و در وی افکند و بر آتش نهد تا بجوشد. پس بپالاید و قدری زاج در وی کند و بکار برد. و این نوع مرکب را نشاستجی گویند، و نوع اول را صمغی.

'Acquire wheat starch and place in a copper pot; put on a slow fire until it turns black but don't burn it. Then bray it and place some gallnuts in water to soften it. Add the gall water to the starch and boil it in water. Afterwards filter it and add a bit of vitriol to the liquid and use. This type of ink is known as "starch ink (*Morakkab-e Nešāstaji*)" while the other kind is known as gum ink (*Morakkab-e Ṣāmḡi*).

R4—Recipe from the treatise *Majmu' al-Ṣanāye'* by Anonymmous (16th C):

صبر نشاسته بستانند و در دیگ کند از مس، و بر آتش نرم بریان کنند تا سیاه شود و بگذارند تا بسوزد، آنگاه بپالایند و قدری زاک بر وی افکنند و در شیشه کنند و بکار برند.

Find aloe starch and place it in a copper pot and heat it on a steady fire till it darkens and burns. Now filter it (after adding water), add vitriol, pour in a glass bottle and use.

R5—Resāle dar bayān-e Tariqe-ye Sāk̲tan-e Morakkab-e Alvān va Kağad̲hā-ye Alvān by Anonymmous (16th C):

نوع دیگر آسانتر است که نشاسته پاک را در تاوه آهنین کند و بر سر آتش بگذارد تا آتش در وی اثر بکند، پاک بسوزد و آنگاه با گلاب حل کند، مداد براق روان شود.

An easier way of making ink is to place some pure and clean starch in an iron pan and put on a fire until it burns, and the pan is in flames. Afterwards, dissolve it in rose water and the result would be a shiny and fluent ink.

R6 Recipe from the treatise *Golestān-e Honar* by Qāżi Aḥmad Qomi (16th C):

بستاند نشاستج گندم و در دیگ مسین کند ، و بر آتش نرم بریان کند تا سیاه گردد، و نگاه دارد تا نسوزد. پس بساید و قدری مازوج در آب کند تا نرم شود و بگذارد تا صاف شود، در میان نشاستج ریزد و و در وی افکند و بر آتش نهد تا بجوشد. پس بپالاید و قدری زاج در وی کند و بکار برد. و این نوع مرکب را نشاستجی گویند، و نوع اول را صمغی.

'Acquire wheat starch and place in a copper pot; put on a slow fire until it turns black but don't burn it. Then bray it and place some gallnuts in water to soften it. Add the gall water to the starch and boil it in water. Afterwards filter it and add a bit of vitriol to the liquid and use. This type of ink is known as "starch ink (*Morakkab-e Nešāstaji*)" while the other kind is known as gum ink (*Morakkab-e* Ṣāmḡi).

It seems like Qāżi Aḥmad took this recipe from the treatise *Resāle dar Bayān-e Ḵaṭ va Morakkab va Kaḡaḏ va Sāḵtan-e Ranghā* (15th C) since it is the exact same recipe as R3.

4.2 Coloured Inks

Compared to black inks, there are far fewer treatises dedicated to coloured, and other types of ink. This could be due to the process being similar to making pigments or extracting dyes. Furthermore, traditionally, the main writing matter used were black and the purpose of coloured inks were not for the main body of text.

The main treatises containing such recipes are from the treatises *Resāle-ye Buqalamun* by Muhammad Šāh ibn Aḥmad Šāh al-Wali al Bahmani (15th C) which is the Persian translation of sections of the Arabic Treatise Umdatul Kuttāb by Ibn Bādis Ṣunhāji (11th C) dedicated on coloured ink, *Majmu ʻ al-Ṣanāye ʻ* (Anonymmous, 16th C) and *Kašf al-Ṣanāye ʻ* by ʻAli Ḥosseini (19th C).

Resāle-ye Buqalamun separates coloured writing matter into coloured inks, and coloured liqs, based on Ibn Bādis' original writing.

Ḥosseini's treatises contains recipes from *Majmu ʻ al-Ṣanāye ʻ*, in addition to ink recipes he translated from European and Indian sources. Due to this, the treatise is infamous for being one of the hardest Persian treatises to decipher and comprehend, as the names of many ingredients given, are English, French, or Hindi words attempted to be written in Persian alphabets, and many of the measurements are English units of measurement as well such as ounces and grains.

4.2.1 White Inks

Table 4.7 White ink: ingredient and measurements

White ink	Primary ingredients	Parts	Amount	Secondary ingredients	Parts	Amount
R1 (15th C)	Gall nuts	–	–	–	–	–
	White starch	–	–	–	–	–
	Gum Arabic	–	–	–	–	–
	Water	–	–	–	–	–

R1—Recipe from the treatise *Resāle-ye Buqalamun* by Bahmani (15th C):

در ساختن مداد سفید بستانند مازو و نیم کوفته کنند و بریزند برو آن قدر آب که او را بپوشاند و بگذارند یک ساعت تا قوت او به آب آمیخته شود. پس بستانند نشاسته سفید سائیده ریخته و بیامیزند به آب مازو و بسایند خوب تا یکی شوند و بگذارند تا صاف شود. آن صاف بردارند و درد بگذارند. پس بستانند صمغ سائیده و حل کنند به آب که از نشاسته گرفته بودند و به آن درد گذاشته بیامیزند و بجنبانند و بگذارند. و هرگاه که خواهند به عمل آرند، بجنبانند و بنویسند.

White ink: Gather gall nuts and pound and break them and add water to cover the gall nuts and allow for an hour the essence to be extracted. Gather white starch, beat to powder, and add to gall water, bray well and filter; separate the solid substances. Then gather gum, beat to powder, and add to filtered starch water and mixed well. In the final step, add the filtered-out material, mix and shake well. Before usage, shake the ink.

4.2.2 *Red Inks*

Table 4.8 Red ink: ingredient and measurements

Red inks	Primary ingredients	Parts	Amount	Secondary ingredients	Parts	Amount
R1 (12th C)	Gallnuts	–	75 g	–	–	–
	Madder	–	75 g	–	–	–
	Gum Arabic	–	75 g	–	–	–
	Water	–	12 kg	–	–	–
R2 (15th C)	Gallnut extract	–	–	–	–	–
	Cinnabar	–	–	–	–	–
	Gum Arabic	–	–	–	–	–
	Water	–	–	–	–	–
R3 (15th C)	Red poppy	–	–	–	–	–
	Myrtle extract	–	–	–	–	–
	Gum Arabic	–	30 g	–	–	–
	Water	–	–	–	–	–
R4 (15th C)	Gall nut extract	–	389.76 g	–	–	–
	Orpiment	–	75 g	–	–	–
R5 (15th C)	Cinnabar	–	–	–	–	–
	Gall nut extract	–	–	–	–	–
	Gum Arabic	–	–	–	–	–
R6 (15th C)	Sumac extract	–	389.76 g	–	–	–
	Gum Arabic	–	162.4 g	–	–	–
	Vitriol	–	–	–	–	–
R7 (15th C)	Red mallow	–	32.48 g	–	–	–
	Lac	–	–	–	–	–
R8 (15th C)	Red poppy	–	–	–	–	–
	Grape vinegar	–	–	–	–	–
	Gum Arabic	–	–	–	–	–
R9 (15th C)	Red poppy	–	–	–	–	–
	Myrtle extract	–	15 g	–	–	–
	Gum Arabic	–	15 g	–	–	–
	Water	–	–	–	–	–

(continued)

Table 4.8 (continued)

Red inks	Primary ingredients	Parts	Amount	Secondary ingredients	Parts	Amount
R10 (16th C)	Madder Water	– –	– –	– –	– –	– –
R11 (19th C)	Brazilwood Vinegar Rock Alum	– 1 bottle	113.39 g – 85.4 g	– – –	– – –	– – –
R12 (19th C)	Brazilwood Rock alum Cream of tartar Gum Arabic Water	2 ½ ½ ½ 16	– – – – –	– – – – –	– – – – –	– – – – –
R13 (19th C)	Vermilion Iron sulphate Oil	1 1 –	– – –	– – –	– – –	– – –
R14 (19th C)	Soda potash Carbonate of ammonia Aluminium oxalate Cochineal	1 1 ½ –	– – – –	– – – –	– – – –	– – – –
R15 (19th C)	Indian catechu Alkaline ash Aloe	1 ½ ¼	– – –	– – –	– – –	– – –

R1—Recipe from the treatise *Resāle-ye Buqalamun* by Bahmani (15th C):

اگر خواهد که از یک ستیر مازوی یک من مداد کند بستاند یک ستیر روغناس و یک ستیر صمغ عربی و بچهار من آب وی را بجوشاند تا به دو من باز آید. وزان پس فرو گیرد تا سرد شود و به کار برد.

If you want to make 1 *Mann* (3 kg) ink with one *Sir* (75 g) gall nuts, purchase 1 *Sir* (75 g) madder, and 1 *Sir* (75 g) gum Arabic and boil these in 4 *Mann* (12 kg) of water until it is reduced to 2 *Mann* (6 kg). Afterwards remove from fire, allow to cool, and use as desired.

R2—Recipe from the treatise *Resāle-ye Buqalamun* by Bahmani (15th C):

در ساختن مداد سرخ بستانند از آب مازو آن قدر که از برای مداد سفید گرفته بودند و بستانند زنجفر شسته. وصفت شستن او آن است که آن را خرد کنند و در ظرفی کنند و آب برو ریزند و بگذارند تا در ته آب نشیند و آب ازو بریزند. چند بار همچنین کنند. پس خشک کنند و بسایند نرم نرم و بیامیزند به آن آب مازو که نگاه داشته بودند و بگذارند یک لحظه. پس بستانند صمغ عربی و حل کنند به آب و بندازند بر آن و بزنند سخت، پس به آن بنویسند.

Red ink: Gather gall water, the same amount you used in making white ink (White Ink—R1), as well as washed cinnabar; and the way to wash it is to break to pieces, place in a bowl, add water and allow the cinnabar to sink to the bottom, then remove the water. Do this several times. Allow the cinnabar to dry, then beat to powder and slowly mix it with the gall water. Finally gather gum Arabic, dissolve in water, and add to the mixture and bray well and use as desired.

R3—Recipe from the treatise *Resāle-ye Buqalamun* by Bahmani (15th C):

در ساختن مداد سرخ تیره بستانند برگ گل که آن را شقایق نعمان را می گویند، گاهی که سخت سرخ باشد. و پاک کنند از آن سیاه ها که دروست و در دیگی کنند با اندک آب و بر سر آتش نهند تا رنگ ازو بیرون آید و به آب آمیخته شود. پس فرود آرند و صاف کنند و بیندازند بر آن، قدر چهاریک آن آب مورد و دو درم صمغ عربی و بنویسند.

Dark red ink: Acquire dark red poppy (Shaqāyeq-e Na‘mān) and clean the dark matter off, then place in a pot with some water and place on fire to allow the colour to be extracted and mixed with water. Afterwards filter and add a quarter of this weight myrtle extract, and 2 *Dirams* (30 g) gum Arabic and write with it.

R4—Recipe from the treatise *Resāle-ye Buqalamun* by Bahmani (15th C):

در ساختن مداد میگون بستانند قدری از مازو و خرد کنند و بریزند برو پنج برابر او آب و بجوشانند و فرود آرند تا سرد شود و صاف کنند و بندازند بر هر یک رطل پنج درم زرنیخ واستعمال کنند.

Wine coloured (Mey-rang) ink: Gather gall nuts, break them, and place in a container and add 5 times the amount of that water, and set to boil. Then remove from heat, allow to cool, and filter. For each *Raṭl* (389.76 g) of gall water, add 5 *Deram* (75 g) orpiment, and use.

R5—Recipe from the treatise *Resāle-ye Buqalamun* by Bahmani (15th C):

در ساختن مداد یاقوت رنگ بستانند زنجفر رمانی خوب و بسایند تا چون مرهم شود. آنگاه با آب مازوی سفید بزنند. پس بگذارند یک لحظه. پس بندازند درو قدری صمغ عربی حل کرده و بنویسند که در غایت خوبی است.

Ruby (Yāqut) Ink: Acquire pomegranate coloured cinnabar and grind well until ointment like. Then add white gall nut extract and allow to rest for a moment. Finally add to it gum Arabic, mix and write with it for it is a great ink.

R6—Recipe from the treatise *Resāle-ye Buqalamun* by Bahmani (15th C):

مداد سماق رنگ بستانند نیم رطل سماق و بندازند بروسه رطل آب و در آفتاب نهند تا رنگ به آب دهد. آنگاه در هم بمالند و پنج روز دیگر در آفتاب نهند و بندازند و بر هر یک رطل پنج وقیه صمغ عربی در هر روز یک وقیه و بگذارند تا درو بگدازد و بستانند زاج آن قدر که خواهند و بسیار نه که می سوزانند مرکب را.

Sumac (Somāq) ink: Gather ½ *Raṭl* (194.88 g) sumac and add to it 3 *Raṭl* (1.17 kg) water and place under sunlight for the colour to be extracted. Then mix and leave for another 5 days and filter while adding gum Arabic in a manner that for each *Raṭl* (389.76 g) of the sumac water, 5 *Waqiye* (162.4 g) of gum Arabic in total should be added; 1 *Waqiye* (32.48 g) each day. Once the gum Arabic is dissolved well, add as much vitriol as desired but not too much for it shall burn the ink.

R7—Recipe from the treatise *Resāle-ye Buqalamun* by Bahmani (15th C):

در ساختن مداد ریحان رنگ بستانند خیری سرخ قدر یک وقیه و بندازند در هاون و بکوبند خوب و صاف کنند در ظرفی و در ظرف آبگینه کنند و قدری از لاک داخل کنند که خوب است.

Basil (Riyhān) ink[7]: Gather 1 *Waqiye* (32.48 g) red mallow (Ḵeiri-e Sorḵ) and beat to in a mortar, then filter and place in a glass container and add some lac which will make it beautiful.

R8—Recipe from the treatise *Resāle-ye Buqalamun* by Bahmani (15th C):

در ساختن مداد خوب بستانند برگ شقایق نعمان، گاهی که سخت سرخ شده باشد. بکوبند خوب و بگذارند در سرکه انگوری و بر سر آتش نهند و بندازند برو قدری صمغ، پس بنویسند.

Good (Ḵub) ink: Acquire red poppy (Shaqāyeq-e Na'mān) when it is dark red. Beat well and soak in grape vinegar, place on fire and add some gum, and write with it.

R9—Recipe from the treatise *Resāle-ye* Buqalamun by Bahmani (15th C):

در ساختن مداد دیگر که هم از شقایق می سازند و او را برمان میگویند بستانند گل را و دورکنند دنباله های او را و بجوشانند تا بیرون آید رنگ ازو. پس فرو دارند و بندازند برو از آب مورد قدر یک درم و از صمغ عربی یک درم. دیگر بجوشانند تا آب او کم شود و بنویسند.

Birmān Ink: Another recipe that uses red poppy (Shaqāyeq-e Na'mān) and is called 'Birmān'. Acquire the flowers and remove the stems, then boil them to extract the colour. Then remove from heat and add to it 1 *Deram* (15 g) myrtle extract and 1 *Deram* (15 g) gum Arabic. Mix and set to boil until reduced in weight, and use.

R10—Recipe from the treatise *Majmu' al-Ṣanāye'* by Anonymous (16th C):

و اگر روناس سه روز در آب ترکند بعده نویسند به آب او، از هر چیز لطیف تر برآید،والسلام.

Should you soak madder (*Runās*) in water for three days and write with that extract, it will be smoother than any other ink.

R11—Recipe from the treatise *Kašf al-Ṣanāye '* by 'Ali Ḥosseini (19th C)[8, 9]

مداد سرخ رنگ - در یک شیشه سرکه چهار اونس برازل اود[8] ساییده، بیامیزند و بجوشانند تا نصف آن باقی ماند. پس از آن از پارچه صاف نمایند و در آن سه اونس سفوف راک الم[9] بیامیزند و بکار برند.

Making red ink: Add 4 ounces (113.39 g) Brazil wood to a glass container filled with vinegar; boil to half its weight. Afterwards, filter through cloth and add 3 ounces (85.04 g) of powdered rock alum, and use.

Similar recipe but with different proportions can be found in Beasley (1878, p. 358) and Mitchell and Hepworth (1904, p. 114).

R12—**R**ecipe from the treatise *Kašf al-Ṣanāye '* by 'Ali Ḥosseini (19th C)[10]

نوع دیگر- برازل اود ۲ حصه، شب یمانی نیم حصه، د کریم اف تارتار[10] نیم حصه، آب ۱۶ حصه بهم آمیخته، بجوشانند، هرگاه نصف آب باقی بماند در پارچه صاف کرده، نیم حصه صمغ اضافه نموده بکاربرند.

---

[7] Based on the red colour outcome, it probably refers to red Rubin basil.

[8] Reads: Brazil wood.

[9] Reads: Rock alum.

[10] Reads: The cream of tartar ($KC_4H_5O_6$).

> Another method: 2 portions of Brazil wood, half a portion of alum, and half a portion cream of tartar is added to 16 portions of water. Mix and boil to half its weight, afterwards, filter through cloth and add half portion of gum Arabic.

The exact same recipe and proportions can be found in The Druggist's General Receipt Book by Beasley (1878).

R13—Recipe from the treatise *Kašf al-Ṣanāye* ' by 'Ali Ḥosseini (19th C)[11]

مرکب سرخ برای مهر نمودن - ورملین[11] چهار جزء، سلفت اف ایرن یک جزء، روغن به قدرحاجت و رنگ سرخ که پسند خاطر باشد، بهم آمیزند و بکار برند.

> Red ink for seals: 4-part vermilion, and 1 part iron sulphate is mixed with enough oil to give you the desired colour.

A recipe with same ingredients but different proportions under the name Dr. Smellie's Ink (for marking linen) can be found in Beasley (1878, p. 362).

R14—Recipe from the treatise *Kašf al-Ṣanāye* ' by 'Ali Ḥosseini (19th C)[12, 13, 14, 15, 16]

سرخ بنیت روشنای استیفن[12] اندک کامن سوداپتاس[13] با کار بونیت اف آمونیا[14] گرفته المضاعف هر دو سفوف کرودار کل کم کم در آن اندازند چنانچه جوش آن تمام شود آن را با پارچه صاف کنند و نیم حصه اکزالک الومنیا [15]شامل نمایند و بعد از سرد شدن به قدر کفایت کوچنیل[16] ساییده در آن آمیزند و بعد از بیست و چهار ساعت با پارچه صاف کنند مرکب سرخ شود.

'Stephen's red: get some common soda potash and carbonate of ammonia in equal portions; mix well and boil. Then filter through cloth and add ½ parts of aluminium oxalate. After it is cooled down ground and add desired amount of cochineal. After 24 h, filter through cloth.

Most likely Ḥosseini is referring to Dr Henry Stephen's red ink. Dr Stephen's registered the Stephen's Ink Company in 1832 (Wild 2014; Reissland et al. 2017) which became one of the biggest ink producers of Great Britain.

R15—Recipe from the treatise *Kašf al-Ṣanāye* ' by 'Ali Ḥosseini (19th C)

کاتِ هندی سرخ آبدار خوب یک جزء، سجی خوب که به فارسی شخار گویند نصف جزء، صبر ربع جزء، هریک را جداجدا کوفته، شب یمانی در آب صاف بخیسانند و صبح در ظرف مس یا برنج کرده، اول کات را گرم نمایند، چون بجوشد آب شخار و آب صبر را بتدریج در آن ریزند و با کفچه همه را آمیخته تا به قوام آید و قابل نوشتن گردد، آن وقت آن را با پارچه صاف کنند و بکار برند.

---

[11] Reads: Vermillion.

[12] Reads: Stephen.

[13] Reads: Common Soda potash.

[14] Reads: Carbonate of ammonia.

[15] Reads: Oxalic Alumina. Probably aluminium oxalate ($C_6Al_2O_{12}$).

[16] Reads: Cochineal.

1 part red-ripe Indian catechu (*Kāt-e Hendi*), ½ parts good alkaline ash/potash, and ¼ parts of aloe are each pounded. Soak alum overnight and in the morning, pour into a copper or brass pot. Heat the pot and add the catechu. Once boiled, add the alkaline ash and aloe slowly and mix well until it gets its desired form. Then filter through cloth and use as desired.

4.2.3 *Blue Inks*

Table 4.9 Blue inks: ingredient and measurements

Blue inks	Primary ingredients	Parts	Amount	Secondary ingredients	Parts	Amount
R1 (19th C)	Sulphate of indigo Prussian blue Oxalic acid Pure Water	– – – –	85.04 g 56.69 g 28.34 g 3.78 kg	– – – –	– – – –	– – – –
R2 (19th C)	Chinese blue (Prussian blue) Oxalic acid Warm water	– – –	56.69 g 28.34 g 1 bottle	– –	– – –	– – –
R3 (19th C)	Sulphate of indigo Water	– –	– –	Chalk –	– –	– –
R4 (19th C)	Powdered black ink Iron sulphate Gum	2 2 1	– –	– –	– –	– –
R5 (19th C)	Prussian blue + sulphuric acid Oxalic acid	1 1	– –	– –	– –	– –

R1—Recipe from the treatise *Kašf al-Ṣanāye* ‘ by ‘Ali Ḥosseini (19th C)[17, 18, 19, 20]

مرکب آبی- سلفت آف اندگو[17]سه اونس در یک گلن[18] آب خالص اندازند تا بگدازد و با یک اونس اکزالک اسید[19] و دو اونس عمده پرشین بلو[20] در یک شیشه آب حل کرده و اول از آب پرشین بلو که آهن در آن زیاد است آن را از تیزاب دور کرده از آب باران یا آب قطره شستن لازم است.

Dissolve 3 ounces (85.04 g) sulphate of indigo in a gallon of water and add this to a solution made of 2 ounces (56.69 g) common Prussian blue, and an ounce (28.34 g) of oxalic acid in a bottle. Remember that Prussian blue contains iron, hence it needs to be washed and mixed with rainwater and distilled water; refrain from using harsh water.

17 Reads: sulphate of indigo.

18 Reads: Gallon.

19 Reads: Oxalic acid ($C_2H_2O_4$).

20 Reads: Prussian blue ([Fe(II)(CN)6]4). It could also read Persian blue which is the same pigment Eastaugh (2008).

R2—Recipe from the treatise *Kašf al-Ṣanāye* ‘ by ‘Ali Ḥosseini (19th C)[21]

نوع دیگر، چائیز بلو[21] دو اونس، آب گرم جوشیده یک شیشه، اکزالک اسید یک اونس. اول چائیز را در آب گذارده بعد در آن اسید آمیزند فورا مرکب شود.

Another method: 2 ounces (56.69 g) Chinese blue, bottle of preboiled warm water, and 1 ounce (28.34 g) oxalic acid. Place Chinese blue in the water, afterwards add the acid; it will turn to ink immediately.

R3—Recipe from the treatise *Kašf al-Ṣanāye* ‘ by ‘Ali Ḥosseini (19th C)[22]

مرکب نیلگون - سلفت آف اندگو را در آب مثل شربت گداخته، مرکب نمایند، و اگر کم رنگ خواهند اندکی چاک[22] در آن بیامیزند.

Indigo ink: Dissolve sulphate of indigo in water as you would with cordial (syrup), and should you want a lighter coloured ink, add some chalk.

Beasley gives a similar recipe under the category of Colours for Liqueurs: "Pink is given by cochineal; yellow by saffron or safflower; violet by litmus; blue by sulphate of indigo, saturated with chalk" (1878, p. 324).

R4—Recipe from the treatise *Kašf al-Ṣanāye* ‘ by ‘Ali Ḥosseini (19th C)

سفوف مداد اسود و سلفت اف ایرن دو جزء، صمغ یک جزء، هر دو را یکجا ساییده، سفوف سازند و یک اونس او را در کاغذ بپیچند. و از یک اونس این سفوف نیم شیشه درست خواهد شد.

2 parts powdered black ink and iron sulphate, and one part gum Arabic are beaten to a powder and wrap an ounce of this in paper. Half a bottle of ink can be made with 1 ounce of this powder.

R5—Recipe from the treatise *Kašf al-Ṣanāye* ‘ by ‘Ali Ḥosseini (19th C)[23]

بنیت بلو روشنای استیفن کامن پرشین بلو[23]، یعنی رنگ آبی که مال ایران است آن را اول در سلفرک اسید که تیز باشد تر کند و باز چند قطره آب شسته، خشک نمایند. بدین طریق که آن را در ر«اکزالک اسید» خوب می گذارند بعد حصه «پرشین بلو» در یک حصه «اگزالک اسید» تر کند و آب به قدر کفاف ریخته آنگاه بکار برند.

Stephen's common Prussian blue, meaning the blue colour that is from Iran: Dip Prussian blue in strong sulphuric acid, then wash with few drops of water and allow to dry. Afterwards soak 1 part of it, in 1-part oxalic acid and add enough water and use as desired.

As previously mentioned, Persian and Prussian is written similarly in this treatise which is why Ḥosseini mistakenly refers to it as blue from Iran.

---

[21] Reads: Čaiz blu which is most likely Chinese blue. According to Eastaugh (2008) it could refer to both Prussian blue, and Han blue (BaCuSi4O10).

[22] Reads: Chalk.

[23] Reads: Stephen common Persian blue.

4.2.4 Green Inks

Table 4.10 Green inks: ingredient and measurements

Green ink	Primary ingredients	Parts	Amount	Secondary ingredients	Parts	Amount
R1 (15th C)	White gall nuts	–	–	–	–	–
	Verdigris	–	–	–	–	–
	Gum Arabic	–	–	–	–	–
	Grape vinegar	–	–	–	–	–
	Water	–	–	–	–	–
R2 (15th C)	Gum ammoniac	–	45 g	–	–	–
	Saffron	–	45 g	–	–	–
	Brazilwood extract	–	–	–	–	–
R3 (15th C)	Cinnabar	–	–	–	–	–
	Gall nut extract	–	–	–	–	–
	Gum Arabic	–	–	–	–	–
	liquid	–	–	–	–	–
	Red lac water	–	–	–	–	–
	Water					
R4 (15th C)	Indigo	–	45 g	–	–	–
	Verdigris	–	15 g	–	–	–
	Water	–	–	–	–	–
R5 (16th C)	Myrtle extract	1	–	–	–	–
	Onion extract	1	–	–	–	–
	Turmeric	1	–	–	–	–
	Madder	1	–	–	–	–
	Alkaline ash (potash)	1	–	–	–	–
R6 (19th C)	Gall nuts	15	–	–	–	–
	Vinegar	1	–	–	–	–
	Water	200	–	–	–	–
	Iron sulphate	5	–	–	–	–
	Iron filings	4	–	–	–	–
	Indigo + sulphuric acid	–	416.4 g	–	–	–
R7 (19th C)	Acetonitrile of	–	–	–	–	–
	chrome	–	–	–	–	–
	Water					
R8 (19th C)	Blue colour	–	–	–	–	–
	(Prussian blue)	–	–	–	–	–
	Yellow colour					
R9 (19th C)	Cream of tartar	1	–	–	–	–
	Verdigris	2	–	–	–	–
	Water	20	–	–	–	–

R1—Recipe from the treatise *Resāle-ye Buqalamun* by Bahmani (15th C):

بستانند مازوی سفید و خرد کنند و آب برو بریزند آنقدر که او را بپوشاند و بگذارند تا قوت مازو در آب آید. پس صاف کنند آب را و بستانند زنگار سبز آنقدر که خواهند و نرم بسایند و بریزند برو سرکه انگوری آن قدر که او را خمیر کند و بگذارند تا خشک شود. دیگر بسایند نرم و بریزند برو آب مازو و بزنند به چوبی تا خوب آمیخته شود. پس بندازند برو صمغ عربي سائیده آن قدر که خواهند و بجنبانند تا آمیخته شود و بنویسند. و در نسخه دیگر این است که بر بالای آجری بگذارند تا خشک شود که این اصل عمل است.

Green Ink: Gather white gall nuts, break them, and add enough water to cover; until the essence of gall is extracted, then filter. Get desired amount of Verdigris (*Zangār*), pound it while adding grape vinegar until a paste is made; allow to dry. Then beat some more while adding gall water and mix well with a stick. Next add desired gum Arabic and mix well. In another treatise it is mentioned to allow to dry on a brick.

R2—Recipe from the treatise *Resāle-ye Buqalamun* by Bahmani (15th C):

در ساختن مداد سبز بستانند سه درم وشق و بخوابند یک شبانروز در آب بقم. آنگاه خمیر کنند به انگشت در همان آب و صاف کنند. پس بندازند برو سه درم زعفران که رنگ او می گردد به رنگ زر، بلکه خوب تر.

Green ink: Gather 3 *Deram* (45 g) gum Ammoniac[24](Woshaq) and soak them in Brazilwood (*Baqam*[25]) extract for one day and night. Then use your fingers to make a paste in the same container, and filter. Then add to it 3 *Deram* (45 g) saffron and the colour would be that of gold, and even better.

R3—Recipe from the treatise *Resāle-ye Buqalamun* by Bahmani (15th C):

در ساختن مدادی پسته رنگ بستانند زنجفر رمانی و بجوشانند و بسایند مانند مرهم و با آب مازوی کوفته حل کرده، در هم کنند و بگذارند لحظه ای. پس به آب صمغ حل کرده، بزنند. دیگر به آب لاک سرخ کنند، بزنند بسیار و استعمال کنند.

Pistachio coloured (*Peste Rang*) ink: Acquire pomegranate/ruby coloured cinnabar, boil it and grind it until ointment like (Marham), then mix it with gall nut water and allow to rest for a moment. Then mix it with gum water and finally add a lot of red lac water, mix well, and use.

R4—Recipe from the treatise *Resāle-ye Buqalamun* by Bahmani (15th C):

در ساختن مداد خوب بستانند سه درم نیل و بسایند خوب به آب گرم تا چون مرهم شود. پس بندازند برو یک درم زنگار و درهم کنند تا سبز شود و به آن بنویسند که خوب است.

Good ink: Gather 3 *Deram* (45 g) indigo and grind well with warm water until ointment like. Then add 1 *Deram* (15 g) Verdigris, and mix until green, and write for it is a good ink.

---

24 Steingass (2004, p.1469).

25 According to Steingass (2004), Baqam is Brazilwood while purple Baqam is logwood, and red Baqam is sapanwood.

R5—Recipe from the treatise *Majmu ' al-Ṣanāye '* by Anonymous (16th C):

اگر آب مازو و پیاز و زردچوبه ونیل و روناس و اشخار یعنی ساجی، از هریک جزوی جمع کند و به آن بنویسد خطی سبز از هر چیز خوبتر آید.

If you mix some myrtle extract, onion extract, turmeric, indigo, madder (*runās*) as well as potash (*Aškār*) and write with it, it will give the best green coloured ink.

R6—Recipe from the treatise *Kašf al-Ṣanāye '* by 'Ali Ḥosseini (19th C)[26, 27, 28]

مرکب سبز و سیاه - مازوی جوکوب پانزده حصه در آب خالص دویست جزء تا یک ساعت بجوشانند بعد در پارچه صاف نموده سلفت اف ایرن ۵ جزء، براده آهن 4 جزء در آن مخلوط سازند و نیم پاینت[26] سلوشن[27] نیل سرمه سا که با سه حصه سلفرک اسید[28] که خوب ساخته باشند. این مرکب سبز نوشته می شود و چون خشک می گردد بسیار روان می باشد.

Green-black ink: Boil 15 parts powdered gall nuts in 200 parts of water for an hour, then filter through cloth. Add 5 parts of iron sulphate, 4 parts of iron filings, and half a pint (473.18 g) of the solution of indigo mixed with sulphuric acid. This is a green ink that is extremely smooth once dried.

This recipe is somewhat similar to a recipe by Mitchel and Hepworth (1904: 95) but with different proportions, and the addition of chalk.

R7—Recipe from the treatise *Kašf al-Ṣanāye '* by 'Ali Ḥosseini (19th C)[29]

نوع دیگر کلسنائید[29] استیوناتمیریت اف کروم به قدر حاجت به آب آمیخته، بنویسند.

Another method: Add desired amount of calcinated acetonitrile of chrome to water and use.

R8—Recipe from the treatise *Kašf al-Ṣanāye '* by 'Ali Ḥosseini (19th C)[30]

نوع دیگر بلو ورنگزرد[30] به قدر حاجت ریخته، بنویسند.

Mix blue, and yellow to the desired amount, and use.

R9—Recipe from the treatise *Kašf al-Ṣanāye '* by 'Ali Ḥosseini (19th C)

کریم اف تارتر یک جزء، زنگار دو جزء، آب خالص بیست جزء بهم آمیخته، بجوشانند و بکار برند.

[26] Reads: Pint.

[27] Reads Solution.

[28] Reads: Sulphuric acid (H_2SO_4).

[29] Reads: Kalsenā'id which is most likely 'calcinated'.

[30] Reads: "بلو و رنگ زرد" - blue and the colour yellow. Since blue is written in transliteration form and not the translation, it most probably means Prussian blue ([Fe(II)(CN)6]4), which has been seen as transliteration in other recipes.

Mix 1 part cream of tartar and 2 parts Verdigris in 20 parts pure water. Then boil and use as desired.

A very similar recipe is found in books by Beasley Beasley (1878, p. 360) and Mitchell and Hepworth (1904, p. 114).

4.2.5 *Yellow Inks*

Table 4.11 Yellow inks: ingredients and measurements

Yellow inks	Primary ingredients	Parts	Amount	Secondary ingredients	Parts	Amount
R1 (15th C)	Yellow orpiment	–	27.84 g	Bran	–	–
	Starch	–	18.56 g	–	–	–
	Gall water	–	–	–	–	–
R2 (19th C)	Lead ceruse	2	–	–	–	–
	Minium	1	–	–	–	–
	Vinegar	–	–	–	–	–
	White gall nut extract	–	–	–	–	–
	Gum Arabic	–	–	–	–	–

R1—Recipe from the treatise *Resāle-ye Buqalamun* by Bahmani (15th C):

در ساختن مداد زرد - بستانند آب مازو آن قدر که زنگار استده بودند و به جای زنگار، زرنیخ زرد بندازند و درینجا سرکه نکنند، ولیکن پارهای نشاسته بندازند و به آب مازو بزنند تا آمیخته شود و اگر پاره ای آب سبوس بندازند خوب تر باشد.

Yellow ink: Prepare gall water and acquire yellow orpiment (Zarnik̲-e Zard), in the same amount you used with Verdigris in the previous recipe (Green ink—R1). Don't use vinegar but instead, add starch, and mix with the gal water; and should you add some bran (Sabus), it shall be better.

R2—Recipe from the treatise *Kašf al-Ṣanāye* ' by 'Ali Ḥosseini (19th C)

در ساختن مداد گلرنگ بستانند اسفیداج ارزیز دو جزو و سرنج یک جزو و خمیر سازند به سرکه و در دیگی کنند که به گل حکمت گرفته باشند و در کوره بالائین آبگینه گران نهند سه شبانروز. پس بیرون آورند و بسایند و بریزند بر آن اندک آب مازوی سفید و در آن اندازند قدری صمغ عربی و بنویسند.

Safflower (Golrang[31]) ink: Gather 2 parts lead ceruse (Esfidāj-e Arziz) and 1-part minium (Soranj) and make a paste with vinegar. Then place in a pot and cover with mud cement (Gel-e Ḥekmat), place in furnace oven, and cover with heavy glass. After three days and nights, remove, pulverize, and add some white gall water and gum Arabic, and use as desired.

[31] The term Golrang can refer to 'colour of flower' (rose) or 'colour-flower' which is another term for safflower (Steingass, 2004).

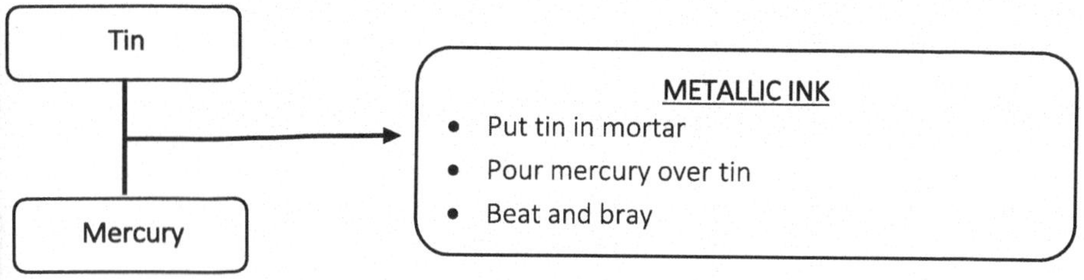

Fig. 4.11 Flowchart guide for constructing Metallic ink

Table 4.15 Metallic ink: ingredients and measurements

Metallic inks	Primary ingredients	Parts	Amount	Secondary ingredients	Parts	Amount
R1 (15th C)	Tin Quicksilver (mercury) Gum Arabic Water	1 1 – –	– – – –	– – – –	– – – –	– – – –
R2 (16th C)	Tin Quicksilver (mercury) Gum Arabic Water	1 1 – –	– – – –	– – – –	– – – –	– – – –

4.2.9 *Metallic Ink*

One recipe was found that mentions the use of two metals to create an ink (Fig. 4.11; Table 4.15).

R1—Recipe from the treatise *Resāle dar Bayān-e Ḵaṭ va Morakkab va Kāgaḏ va Sāḵtan-e Ranghā* by Anonymous (15th C) (Fig 4.12):

قلعی بگیرد و سیماب، هر یکی را در برابر یکدیگر، و قلعی را بگذاد و سیماب بر وی ریزد و صلایه کند. چنان که نرم شود و آب صمغ عربی در وی ریزد و کتابت کند و بر پشت مهره زند.

Find quicksilver (mercury) (*Simāb*) and tin (*Qal'*), in equal portions. Pour quicksilver on tin and bray well. Once softened, add liquid gum Arabic to the metal mixture and bray. Now write with it and rub it on the back of seals.

R2 Recipe from the treatise *Golestān-e Honar* by Qāżi Aḥmad Qomi (16th C):

بگیرد قلع و سیماب هریکی را برابر یکدیگر و قلعی را بگدازد و سیماب بر وی ریزد وصلایه کند چنانچه نرم شود و آب صمغ عربی در وی ریزد و کتابت کند و بر پشت مهره بمالد .

Find quicksilver (mercury) (*Simāb*) and tin (*Qal'*), in equal portions. Pour quicksilver on tin and bray well. Once softened, add liquid gum Arabic to the metal mixture and bray. Now write with it and rub it on the back of seals.

The exact same recipe as R1 was used by Qāżi Aḥmad a century later.

4.2.6 *Red-Green Inks*

Table 4.12 Red-green ink: ingredient and measurements

Red-green inks	Primary ingredients	Parts	Amount	Secondary ingredients	Parts	Amount
R1 (15th C)	Honey Talc Green vitriol Gum Arabic	1 1 1 –	– – – 300 g	– – – –	– – – –	– – – –

R1—Recipe from the treatise Resāle-ye *Buqalamun* by Bahmani (15th C):

در ساختن‌مداد سرخ سبز بستانند یک جزو عسل و یک جزو طلق و یک جزو زاج سبز و با همدیگر بسایند و در آفتاب نهند بیست روز و هر روز یک درم صمغ برو اندازند و بجنبانند سخت تا آن صمغ بگدازد و به آن بنویسند.

Red-green ink: Take a portion of honey, a portion of talc, and a portion of green vitriol. Grind together and place under sunlight for 20 days, and each day add 1 *Deram* (15 g) of gum and shake well until dissolved and write with it.

4.2.7 *Black & White Inks*

Table 4.13 Black and white inks: ingredient and measurements

Black and white Inks	Primary ingredients	Parts	Amount	Secondary ingredients	Parts	Amount
R1 (15th C)	Pumice stone Antimony collyrium Marcasite Litharge Vitriol White ceruse Gall nuts Gum Arabic Starch Water	– – – – – – – – – –	60 g 45 g 45 g 15 g 60 g 30 g 75 g – – –	– – – – – – – – – –	– – – – – – – – – –	– – – – – – – – – –

R1—Recipe from the treatise *Resāle-ye Buqalamun* by Bahmani (15th C):

در ساختن‌مداد که بر سیاه، سفید نماید و بر سفید، سیاه؛ مدادی عجیب است بستانند قلیه چهار درم و بندازند برو نیم رطل آب و بجنبانند هفت روز و هرگاه که آب چیزی کم شود همان قدر اضافه کنند و بجنبانند. چون هفت روز تمام شود صاف کنند آب او را بر سه درم سرمه عملی ، که آن سرمه دراهم است و سه درم مرقشیشای سائیده و یک درم مردارسنج بکوبند هریک جداگانه و با هم در هاونی اندازند و اضافه کنند به او چهار درم زاج و دو درم اسفیداج رصاص. پس بسایند همه را نرم و بریزند برو سه وقیه آب و بگذارند پنج روز. پس بستانند از قلیه یک وقیه و از سرمه یک وقیه و از مازوی خرد کرده پنج درم و بر آتش دارند تا ثلث آب کم شود و اگر با مازو براده آهن و نشاسته باشد خوب است. پس در هم کنند آن دو آب را و بندازند قدری صمغ عربی و نشاسته و به آن بنویسند.

Ink that would appear white on black paper, and black on white paper for it is a unique ink: Acquire 4 *Deram* (60 g) pumice stone (*Qaliye*),[32] add to it 1 *Raṭl* (389.76 g) of water and shake for 7 days and what amount of water is lost, add more and shake. After the 7th day, filter it. Meanwhile gather 3 *Deram* (45 g) antimony collyrium (*Sorme-ye 'Amali*) and 3 *Deram* (45 g) beaten marcasite and a *Deram* (15 g) of litharge (Mardarsānj).

Beat each to powder separately, then add all to a mortar, and add to it 4 *Deram* (60 g) vitriol, and 2 *Deram* (30 g) white ceruse (of lead or tin) and beat to powder and bray well, then add to it 3 *Waqiye* (97.44 g) of water and allow to rest for 5 days.

Now gather from the mixtures, 1 *Waqiye* (32.48 g) pumice stone, 1 *Waqiye* (32.48 g) antimony, 5 *Deram* (75 g) beaten gall nuts and boil together until a third of the liquid is reduced. And if the gall nut water contains vitriol and starch, it shall be better. Now mix the two solutions together, then add some gum Arabic and some starch, and use as desired.

4.2.8 Golden Ink

Table 4.14 Golden ink: ingredients and measurements

Golden inks	Primary ingredients	Parts	Amount	Secondary ingredients	Parts	Amount
R1 (12th C)	Tin Saffron Gum Arabic Water	– – – –	150 g 45 g 30 g 389.76 g	– – – –	– – – –	– – – –
R2 (15th C)	White ceruse Vitriol Gum Arabic liquid	– – –	27.84 g 18.56 g –	– – –	– – –	– – –
R3 (19th C)	Leaf gold Powdered gold Alcohol Honey Gum Arabic	22 pieces – 30 drops – –	– 14.17 g – 9.5 g 60 g	– – – – –	– – – – –	– – – – –
R4 (19th C)	Orpiment Gold leaf Blue clay Egg yolks	– 45 pieces 3 parts –	36.8 g – – 36.8 g	– – – –	– – – –	– – – –

[32] Steingass (2004, p. 987).

R1—Recipe from the treatise *Bayān al- Ṣanā 'āt* by Teflisi (12th C):

سنگ ارزیز و او را در پاتیله کند و یک رطل بکوچک آب در وی کند و ارزیز را بدو بمالد. وزان پس به
را بجوشاند تا رنگ ارزیز به آب ازو بیرون آید. پس بپالایدش و سه درم سنگ زعفران ریشه با وی بیامیزد
می جوشاند چندتانکه رنگ زعفران جمله بینداید. پس از آن آب را صافی کند و دو درم صمغ عربی کوفته
پس از آنکه سرد شود همی جنباند تا سخت نیکو و پاکیزه گردد و بکار برد.

Acquire 10 *Deram* (150 g) of tin (*Arziz*), place in a pot and a (389.76 g) water and mix them well. Boil it them on a steady flame unti takes the colour of tin. Next filter and add 3 *Derams* (45 g) of saffron s set to boil again until saffron colour is extracted. Filter again and add (30 g) powdered gum Arabic. After it has cooled down mix it well u reaches desired look and consistency and start using it.

R2—Recipe from the treatise *Resāle-ye Buqalamun* by Bahmani (15th

که به او بنویسند مانند طلا بستانند شش مثقال اسفیداج و چهار مثقال زاج سبز. همه در ظرف آبگینه کنند
بگیرند و در کوره بالائین آبگینه گران نهند یک شبانروز. پس بیرون آرند و بریزند برو قدری آب صمغ
ب خوب است.

Gold coloured ink: Gather 6 *Meṯqāl* (27.84 g) white ceruse an (18.56 g) vitriol. Place in a glass container and close the lid tightly. above a strong glass container and place in oven for a night and day. T add to it some gum Arabic liquid and write with it for it is a unique ink

R3—Recipe from the treatise *Kašf al-Ṣanāye* ' by 'Ali Ḥosseini (19th

دو عدد، سفوف طلا نیم اونس، جوهر شراب، یعنی اسپریت سی قطره، شهد خالص سی گرین[33]، صمغ
آب باران یا آب قطره چهار اونس. اول طلا را در شهد و صمغ حل نمایند بعد از آن آب و اسپریت شامل

22 pieces of leaf gold, ½ ounce (14.17 g) powdered gold, 30 dro (*Johar-e Sharāb*) or spirits, 30 grains (1.95 g) of honey, 4 *Derams* Arabic, 4 oz of rainwater or distilled water. First mix and bray the g and gum, then add water and spirit.

R4—Recipe from the treatise *Kašf al-Ṣanāye* ' by 'Ali Ḥosseini (19th

دول کتب و غیره - زرنیخ طبقی چهار توله، ورق طلا چهل و پنج عدد، تباشیر کبود سه هسته، زرده
ه، همه را خوب ساییده اندک آب گرم در آن ریزند و بیامیزند و ساعتی نگاه داشته که ته نشین شود و
ستاده آهسته آهسته بریزند تا یک هفته به آب صمغ پلاس حل نموده بعد در شیشه ریخته چهار روز در
از آن بکار برند.

Golden ink for framing lines of text block and other uses: 4 *Tuleh* ment plates, 45 pieces of leaf gold, 3 parts (*Haste*) of blue clay, 4 egg yolks. Mix and beat well, then add some warm water, mix we for the particles to precipitate for one hour. Then pour pure clean w it slowly and add filtered gum water to it over a week. Now pour in container, place under sunlight for 4 days, and use.

[33] Reads: Grain. The unit of measurement equivalent to 0.065 g.

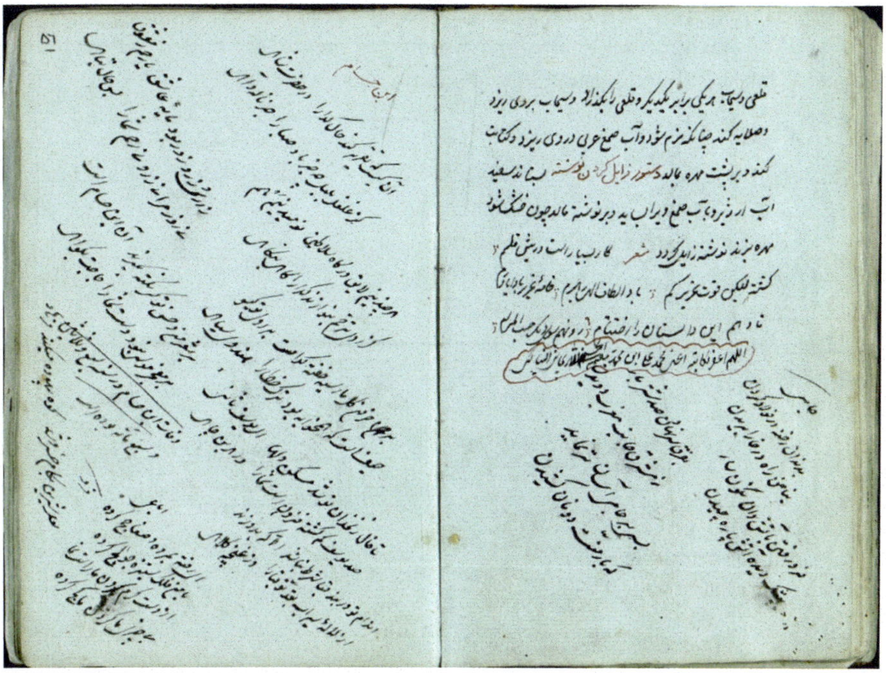

Fig. 4.12 Recipe on Metallic Ink (Right) from the treatise Resāle dar Bayān-e Ḵaṭ va Morakkab va Kāgaḏ va Sāḵtan-e Ranghā by Anonymous (15th C). Manuscript 4283—Tehran University Central Library

4.3 Coloured Liqs

As with Ibn Bādis' original treatises Umdatul Kuttāb (11th C), Bahmani (15th C) in *Resāle-ye Buqalamun* separates coloured writing matter to coloured inks, and coloured *liqeh* or liqs.

4.3.1 White Liqs

Table 4.16 White liqs: ingredient and measurements

White liq	Primary ingredients	Parts	Amount	Secondary ingredients	Parts	Amount
R1 (15th C)	White ceruse Gall nut extract Gum Arabic	– – –	– – –	– – –	– – –	– – –

R1—Recipe from the treatise *Resāle-ye Buqalamun* by Bahmani (15th C):

در ساختن لیقة سفید خام بستانند اسفیداج آن قدر که خواهند و بسایند خوب به آب مازو که درو جوشانیده باشند یک ساعت. پس خشک کنند و بگذارند به آب صمغ آن قدر که خواهند و به آن بنویسند، اگر حق تعالی خواسته باشد.

White coloured liq: Acquire desired amount of white ceruse and pound and grind well for one hour with gall water which was previously boiled. Allow to dry, then add desired amount of gum Arabic solution and write with it.

4.3.2 *Red Liqs*

Table 4.17 Red liqs: ingredient and measurements

Red liqs	Primary ingredients	Parts	Amount	Secondary ingredients	Parts	Amount
R1 (15th C)	Black clay	–	–	–	–	–
	Cinnabar	–	–	–	–	–
	Brazilwood extract	–	–	–	–	–
R2 (15th C)	Burnt copper	1	–	–	–	–
	Yellow orpiment	½	–	–	–	–
	Gum Arabic liquid	–	–	–	–	–
R3 (15th C)	Burnt copper	1	–	–	–	–
	Indigo	1	–	–	–	–
	Gum Arabic liquid	–	–	–	–	–
R4 (15th C)	Talc	1	–	–	–	–
	Red orpiment	1	–	–	–	–
	Brazilwood extract	–	–	–	–	–
R5 (15th C)	Green gall nut	–	–	–	–	–
	Grape vinegar	–	–	–	–	–
	Saffron	–	–	–	–	–
	Gum Arabic	–	–	–	–	–
R6 (15th C)	Green gall nut	–	–	–	–	–
	Grape vinegar	–	–	–	–	–
	Saffron	–	–	–	–	–
	Gum Arabic	–	–	–	–	–

R1—Recipe from the treatise *Resāle-ye Buqalamun* by Bahmani (15th C):

در ساختن لیقه سرخ بستانند از آن خردهای راست کرده آن قدر که خواهند و بسایند نرم و بندازند بر آن از زنجفر شسته رمانی سائیده آن قدر که بس کند. پس در ظرفی خوب پاکیزه کنند و بریزند بر آن از آب بقم تازه آن قدر که او را بپوشاند و بسازند از آن لیقه را و به آن بنویسند.

Red liq: Gather prepared black viscous clay[34] (*Ḵardhā-ye Rāst Karde*) as much as desired and beat and turn to powder. Add washed and powdered pomegranate coloured cinnabar (*Šanjarf-e Romāni*). Pour these in a clean container and add enough freshly made Brazilwood extract to cover, and grind and bray this ink, and write with it.

R2—Recipe from the treatise *Resāle-ye Buqalamun* by Bahmani (15th C):

در ساختن لیقهٔ سخت سرخ بستانند از مس سوخته یک جزو و از زرنیخ زرد نیم جزو و هر یک بکوبند خوب تنها. پس به هم بیامیزند به سائیدن خوب و بگذارند به آب صمغ و بنویسند.

Strong red liq: Acquire 1 part burnt copper and ½ part yellow orpiment and beat each to powder separately. Then pound and mix them together and add gum water and write with it.

[34] Steingass (2004, p.453).

R3—Recipe from the treatise *Resāle-ye Buqalamun* by Bahmani (15th C):

در ساختن لیقه می رنگ بستانند مس سوخته یک جزو و از نیل یک جزو و بکوبند هر یک جداگانه و در ظرفی کنند و بندازند بر ایشان آب صمغ و به آن بنویسند.

> Wine coloured (*Mey-rang*) liq ink: Acquire 1 part burnt copper (*Mes-e Sukte*), and 1 part indigo. Beat each to powder and place in separate containers. Then add to them gum water and write with it.

R4—Recipe from the treatise *Resāle-ye Buqalamun* by Bahmani (15th C):

در ساختن لیقه سرخ سیر بستانند طلق آن قدر که خواهند و خورد کنند به ناخن چین مانند خردل و در جامه ای کنند شفت و بمانند تا آنچه خواهند ازو بیرون آید. پس او را ببیزند به جامه شفت. آنگاه بستانند یک جزو ازو و یک جزو زرنیخ سرخ، بسایند نرم و با هم خلط کنند. پس خمیر کنند به آب صمغ که بعد از این وصف خواهیم کرد و خوش کنند و بردارند و چون خواهند که به او بنویسند قدری ازو بستانند و بگذارند در صدفی به آب صمغ و بنویسند.

> Red liq ink: Acquire Talc as much as desired and cut to small pieces like mustard, with a nail clipper, and squeeze it in in an old rag to extract and remove its essence. Take a part of this and a part of red orpiment, grind and mix, then make a paste by adding gum water that will be described after this and allow to dry. For usage, take some, place in a seashell, add and mix gum Arabic, and set to write.

Preparing gum Arabic is later described as beating to powder 1 *Raṭl* (389.76 g) gum Arabic, adding clear water, boiling on low fire, filtering and cooling down, and finally using.

The recipe adds that for making gold colour ink using this recipe, use yellow orpiment instead of red (Golden liq—R4).

R5—Recipe from the treatise *Resāle-ye Buqalamun* by Bahmani (15th C):

در ساختن لیقه گلناری بستانند مازوی سبز و خورد کنند و بیامیزند با هم چند او سرکه تیز و بگذارند یک لحظه. پس صاف کنند نیکو و بیامیزند با قدری زعفران یا آب زعفران و بجوشانند با او صمغ سائیده و استعمال کنند.

> Pomegranate flower (*Golnāri*) liq: Gather green gall nuts, break them and add many times of its weight, strong vinegar and allow to rest for a moment. Then filter and add to it some saffron or saffron-water, and powdered gum, set to boil and use.

R6—Recipe from the treatise *Resāle-ye Buqalamun* by Bahmani (15th C):

در ساختن لیقه گلناری بستانند مازوی سبز و خرد کنند و بیامیزند به هم چند او سرکه تیز و بگذارند لحظه ای. پس صاف کنند و بیامیزند با او قدری زعفران یا آب زعفران با صمغ سائیده، آنگاه استعمال کنند.

> Pomegranate flower (*Golnāri*) liq ink: Gather green gall nuts, break them and add many times of its weight, strong vinegar and allow to rest for a moment. Then filter and add to it some saffron or saffron-water and powdered gum, then use as desired.

This is the exact same recipe as the previous with the only difference that boiling before usage is not mentioned.

4.3.3 Purple Liqs

Table 4.18 Purple liqs: ingredient and measurements

Purple liqs	Primary ingredients	Parts	Amount	Secondary ingredients	Parts	Amount
R1 (15th C)	Turmeric	–	–	–	–	–
	Saffron	–	150 g	–	–	–
	Myrtle/pome-	–	–	–	–	–
	granate rind	15 g	–	–	–	–
	Gum Arabic	–	–	–	–	–
	Water					

R1—Recipe from the treatise *Resāle-ye Buqalamun* by Bahmani (15th C):

در ساختن لیقة بنفش بستانند قدری زردچوبه و برو ریزند آنقدر آب که او را بپوشاند در ظرفی مسین کوچک و بپزند و فرو آرند و صاف کنند آب او. بستانند ده درم زعفران موی و در آب کنند همچنان که هست و بجوشانند تا آن زمان که پر رنگ کند. آنگاه صاف کنند و بستانند از آب مورد یا از آب پوست انار آن قدر که بخورد. پس بندازند برو یک درم صمغ و بنویسند به آن.

Purple liq: Gather some turmeric, place in a small copper pot and add enough water to cover it and set to boil, then filter it. Acquire 10 *Deram* (150 g) saffron strands and add to the mixture and boil until it changes colour feather. Afterwards filter and add myrtle water or pomegranate rinds as much as it can take. Finally add 1 *Deram* (15 g) gum Arabic, mix, and write with it.

4.3.4 Blue Liqs

Table 4.19 Blue liqs: ingredient and measurements

Blue liqs	Primary ingredients	Parts	Amount	Secondary ingredients	Parts	Amount
R1 (15th C)	Lapis Lazuli	–	–	–	–	–
	Gall nut extract	–	–	–	–	–
	Gum Arabic	–	–	–	–	–
	Water	–	–	–	–	–

R1—Recipe from the treatise *Resāle-ye Buqalamun* by Bahmani (15th C):

در ساختن لیقة لاجوردی بستانند لاجورد خوب آنقدر که خواهند و برو ریزند آن قدر آب که او را بپوشاند. پس بجنبانند و یک شب بگذارند تا به ته بنشیند و آب ازو بریزند. پس برو ریزند آب مازو که صمغ درو خوسانیده باشند و بنویسند.

Lapis lazuli (*Lājvardi*) liq: Acquire desired amount of good quality lapis lazuli, place in a container and add enough water to cover. Next shake and leave over night to allow the lapis to sink to the bottom and decant the water. Next add gall extract with dissolved gum Arabic and write with it.

4.3.5 Green Liqs

Table 4.20 Green liqs: ingredient and measurements

Green liqs	Primary ingredients	Parts	Amount	Secondary ingredients	Parts	Amount
R1 (15th C)	Gall nuts Saffron Gum Arabic Water	– – – –	– – 30 g –	– – – –	– – – –	– – – –
R2 (15th C)	Verdigris Gum Arabic Gall nut extract Grape vinegar	– – – –	– – – –	– – – –	– – – –	– – – –
R3 (15th C)	Verdigris Gum Arabic Vinegar	3 2 –	– – –	– – –	– – –	– – –
R4 (15th C)	Verdigris Gum Arabic Vinegar	3 2 –	– – –	– – –	– – –	– – –
R5 (15th C)	Myrobalan Gall nut extract Gum Arabic	– – –	– – 16.24 g	– – –	– – –	– – –
R6 (15th C)	Turmeric Indigo Gall nut extract Gum Arabic Water	– – – – –	30 g – – – –	– – – – –	– – – – –	– – – – –
R7 (15th C)	Turmeric Saffron Gum Arabic Myrtle/pomegranate rind extract	– – – –	150 g 15 g 30 g	– – – –	– – – –	– – – –
R8 (15th C)	Turmeric Saffron Gum Arabic Myrtle/pomegranate rind extract	– – – –	150 g 15 g 30 g	– – – –	– – – –	– – – –
R9 (15th C)	Verdigris Gum Arabic Gall nut extract Grape vinegar	– – – –	– – – –	– – – –	– – – –	– – – –

R1—Recipe from the treatise *Resāle-ye Buqalamun* by Bahmani (15th C):

در ساختن لیقه سبز بستانند مازو و خرد کنند و برو ریزند آن قدر آب که او را بپوشاند و یک لحظه بگذارند تا آب قدری قوت ازو بستاند. پس صاف کنند و نگاه دارند. دیگر بستانند زعفران خوب آن قدر که خواهند و بشویند به آب همچنان که گذشت. پس بستانند آنچه به ته نشسته باشد از غلیظ آن و بسایند و در ظرفی کنند و برو ریزند آب بسیار و بجنبانند و از آنجا به ظرفی دیگر کنند و به این طریق چند بار عمل کنند تا تمام فارغ شود. پس بگذارد تا به ته نشیند و آب ازو برود و خشک شود. آنگاه بستانند ازو آنچه خواهند و بسایند با قدری آب مازو که نگاه داشته بودند و بندازند در آن دو درم صمغ عربی تا آن قدر که بس کند و همه با هم بسایند و بنویسند.

Green liq: Gather gall nuts and break them to pieces and cover with water and allow to rest so the water gets strengthened by the extract; filter and keep aside. Now gather desired amount of saffron and wash it in the manner previously explained (in other recipes). So, gather the thick sunken saffron, bray well, place in a container and a lot of water, shake and decant the water. Do this a few times until all dirt is removed; allow to dry. Take desired amount of this and grind it along with the gall water that was set aside and add to it 2 *Deram* (30 g) gum Arabic and mix all together and write with it.

R2—Recipe from the treatise *Resāle-ye Buqalamun* by Bahmani (15th C):

در ساختن لیقه سبز بستانند زنگار و بسایند و با هم چند او صمغ عربی سفید و آب مازو و اندکی سرکه و استعمال کنند، اگر حق تعالی خواسته باشد.

Green liq: Gather Verdigris, and many times its weight white gum Arabic and gall extract and grind them along with some grape vinegar and use it God willing.

R3—Recipe from the treatise *Resāle-ye Buqalamun* by Bahmani (15th C):

در ساختن لیقة سبز دیگر بستانند زنگار سه جزو و دو جزو صمغ و بسایند با سرکه نیکو و بیفزایند سرکه آن قدر که خواهند و بنویسند به آن که خوب است.

Another green liq: Gather 3 parts Verdigris, and 2 parts gum Arabic, and grind them with good vinegar as much as desired and write with it.

R4—Recipe from the treatise *Resāle-ye Buqalamun* by Bahmani (15th C):

در ساختن لیقه سبز دیگر بستانند زنگار سه جزو و صمغ دو جزو و بسایند به سرکه انگوری خوب و هر چند سرکه بخورد بدهند و به آن بنویسند.

Another Green Liq: Gather 3 parts Verdigris, and 2 parts gum Arabic, and grind them along with grape vinegar as much as it can take and write with it.

Same recipe as before except this one mentions the use of "grape" vinegar.

R5—Recipe from the treatise *Resāle-ye Buqalamun* by Bahmani (15th C):

در ساختن لیقه سبز بستانند مازو و خرد کنند و بریزند برو آب هلیله با نیم وقیه صمغ عربی. پس بنویسند به آن که خوب است.

Green liq: Gather gall nuts and break them and add to them Myrobalan (*Halile*) water and ½ *Waqiye* (16.24 g) gum Arabic. So, write with it for it's a good ink.

R6—Recipe from the treatise *Resāle-ye Buqalamun* by Bahmani (15th C):

در ساختن لیقة ازرق، بستانند زردچوبه دو درم و بندازند در دیگی مسین و بر وی ریزند آن قدر آب که او را بپوشاند و بپزند آن قدر که پر رنگ کند. پس فرو آرند و صاف کنند و بندازند برو آن قدر نیل که بس کند و بزنند با آب مازو در هم و بندازند برو قدری صمغ سائیده و استعمال کنند آنچه خواهند، ان شاء الله تعالی.

Green (*Azraq*) Liq: Pour 2 *Deram* (30 g) of turmeric in a copper pot and add enough water to cover, and boil to the point that it changes the colour of feathers. Then filter and add enough indigo, and mix it with gall water, and add to it some ground gum Arabic and use as desired.

R7—Recipe from the treatise *Resāle-ye Buqalamun* by Bahmani (15th C):

در ساختن لیقه پسته رنگ بستانند ده درم زردچوبه و بریزند بر آن آنقدر آب که او را بپوشاند در ظرفی مسین کوچک و بپزند آن قدر که اگر یک پر مرغ درو فرود برند رنگین شود. پس فرود آرند و بگذارند تا صاف شود و بستانند یک درم زعفران موی و بندازند همچنان در آب و بجوشانند آن قدر که پر رنگ کند. پس صاف کنند و بستانند آب مورد یا آب پوست انار، هر کدام که خواهند، به قدر آنچه باید بیش نه تا سیاه نشود و بندازند در آن دو درم صمغ کوفته بیخته و بنویسند.

Pistachio coloured (*Peste rang*) liq: Gather 10 *Deram* (150 g) turmeric, place in a small copper pot and add enough water to cover it and set to boil to the point that if a feather is dipped, it will change its colour, then filter it. Acquire 1 *Deram* (15 g) saffron strands and add to the mixture and boil until it changes the colour of feather. Afterwards filter and add myrtle water or pomegranate rinds water whichever is desired, enough amount to not make the colour black. Finally add 2 *Deram* (30 g) powdered gum Arabic, mix, and write with it.

R8—Recipe from the treatise *Resāle-ye Buqalamun* by Bahmani (15th C):

در ساختن لیقه پسته ای بستانند ده درم زردچوبه و برو ریزند آن قدر آب که او را بپوشاند و در دیگ مسین کنند و بپزند تا آن زمان که پر رنگ تواند کرد و پس فرود آرند و بستانند یک درم زعفران موی و در آب اندازند و بجوشانند تا آن زمان که پر رنگ تواند کرد. پس صاف کنند و بسایند به آب مورد یا آب پوست انار هر کدام که خواهند بسیار نه و بندازند درو صمغ عربی کوفته بیخته و بنویسند به آن که خوب است.

Pistachio coloured (*Peste-yi*) liq: Gather 10 *Deram* (150 g) turmeric, add enough water to cover it, place in a small copper pot and a and set to boil to the point that if a feather is dipped, it will change its colour, then filter it. Acquire 1 *Deram* (15 g) saffron strands and add to the mixture and boil until it changes colour of the feather. Afterwards filter and add a lot of myrtle water or pomegranate rinds water whichever is desired and add 2 *Deram* (30 g) beaten gum Arabic, mix, and write with it for it is a good ink.'

The recipe is exactly the same as previous, except for the fact that it mentions using "a lot" of myrtle or pomegranate rind water, and there is no mentioning of the ink turning black if too much is used.

R9—Recipe from the treatise *Resāle-ye Buqalamun* by Bahmani (15th C):

در ساختن لیقه زمردی بستانند زنگار و بسایند با هم چند او صمغ عربی سفید به آب مازو. پس بریزند برو اندکی سرکه و استعمال کنند.

Emerald (*Zommrodi*) liq ink: Gather Verdigris and pulverize it along with many times its weight white gum Arabic, and gall nut extract. Finally add to it some vinegar and use it.

4.3.6 *Yellow Liqs*

Table 4.21 Yellow liqs: ingredient and measurements

Yellow liqs	Primary ingredients	Parts		Amount	Secondary ingredients	Parts	Amount
R1 (15th C)	Yellow orpiment		1	–	–	–	–
	Gum Arabic		–	–	–	–	–
	Water		–	–	–	–	–
R2 (15th C)	Yellow orpiment		–	–	–	–	–
	Gum Arabic		–	–	–	–	–
	Saffron		–	–	–	–	–
	Egg yolk		–	–	–	–	–
	Warm water		–	–	–	–	–

(continued)

Table 40.21 (continued)

Yellow liqs	Primary ingredients	Parts	Amount	Secondary ingredients	Parts	Amount
R3 (15th C)	Yellow orpiment	–	–	–	–	–
	Gall nut extract	–	–	–	–	–
	Gum Arabic	–	–	–	–	–
	Iraqi indigo	–	–	–	–	–
	Leek extract	–	–	–	–	–
R4 (15th C)	Yellow orpiment	1	–	–	–	–
	Saffron	1	–	–	–	–
	Gum Arabic	–	–	–	–	–
	Water	–	–	–	–	–
R5 (15th C)	Lead ceruse	1	–	–	–	–
	Minium	1	–	–	–	–
	Vinegar	–	–	–	–	–
	Gall nut extract	–	–	–	–	–
	Gum Arabic	–	–	–	–	–

R1—Recipe from the treatise *Resāle-ye Buqalamun* by Bahmani (15th C):

در ساختن لیقه سخت زرد بستانند زرنیخ زرد مصفح یک جزو و بسایند با هم چندان او صمغ عربی و در ظرفی پاک کنند و برو ریزند آب صمغ آن قدر که او را بپوشاند و بنویسند.

Strong yellow liq: Acquire 1 part fine yellow orpiment and add many parts of that washed gum Arabic (liquid), to cover, and write with it.

R2—Recipe from the treatise *Resāle-ye Buqalamun* by Bahmani (15th C):

در ساختن لیقه زرد آفتابی بستانند زرنیخ و تر کنند به آب نیم گرم همراه صمغ و زعفران تا همه با هم حل شوند و بیامیزند به آن زرده تخم مرغ و در پرز صوف سفید بکنند و به آن بنویسند.

Sun yellow liq: Gather orpiment and soak it in warm water and add gum and saffron until they are mixed uniformly. Then add egg yolk and pour over white wool cloth[35] (*Porz-e Ṣuf-e Sefid*) and write with it.

R3—Recipe from the treatise *Resāle-ye Buqalamun* by Bahmani (15th C):

لیقه زرد آلو رنگ بستانند زرنیخ زرد آن قدر که خواهند و بسایند به آب مازو و آب صمغ تا نرم شود و خشک کنند و بستانند از ویک جزو و چند شش یک او نیل عراقي وبسایند به آب گندنا و در هم کنند و بگذارند و به آن بنویسند، اگر حق تعالی خواسته باشد.

Apricot coloured liq: Gather yellow orpiment as much as desired, and grind with gall extract and liquid gum Arabic until smooth and fine. Once dried, take one part of this, and a sixth of this amount Iraqi indigo, and grind it along with leek extract (*Āb-e Gandanā*) and mix well and write with it, God willing.

[35] This is the silk, cotton or wool strands known as līq or līqeh that are used in inkwells.

4.2.6 *Red-Green Inks*

Table 4.12 Red-green ink: ingredient and measurements

Red-green inks	Primary ingredients	Parts	Amount	Secondary ingredients	Parts	Amount
R1 (15th C)	Honey Talc Green vitriol Gum Arabic	1 1 1 –	– – – 300 g	– – – –	– – – –	– – – –

R1—Recipe from the treatise Resāle-ye *Buqalamun* by Bahmani (15th C):

در ساختن مداد سرخ سبز بستانند یک جزو عسل و یک جزو طلق و یک جزو زاج سبز و با همدیگر بسایند و در آفتاب نهند بیست روز و هر روز یک درم صمغ برو اندازند و بجنبانند سخت تا آن صمغ بگدازد و به آن بنویسند.

Red-green ink: Take a portion of honey, a portion of talc, and a portion of green vitriol. Grind together and place under sunlight for 20 days, and each day add 1 *Deram* (15 g) of gum and shake well until dissolved and write with it.

4.2.7 *Black & White Inks*

Table 4.13 Black and white inks: ingredient and measurements

Black and white Inks	Primary ingredients	Parts	Amount	Secondary ingredients	Parts	Amount
R1 (15th C)	Pumice stone Antimony collyrium Marcasite Litharge Vitriol White ceruse Gall nuts Gum Arabic Starch Water	– – – – – – – – – –	60 g 45 g 45 g 15 g 60 g 30 g 75 g – – –	– – – – – – – – – –	– – – – – – – – – –	– – – – – – – – – –

R1—Recipe from the treatise *Resāle-ye Buqalamun* by Bahmani (15th C):

در ساختن مداد که بر سیاه، سفید نماید و بر سفید، سیاه؛ مدادی عجیب است بستانند قلیه چهار درم و بندازند برو نیم رطل آب و بجنبانند هفت روز و هرگاه که آب چیزی کم شود همان قدر اضافه کنند و بجنبانند. چون هفت روز تمام شود صاف کنند آب او را بر سه درم سرمه عملی ، که آن سرمه دراهم است و سه درم مرقشیشای سائیده و یک درم مردارسنج بکوبند هریک جداگانه و با هم در هاونی اندازند و اضافه کنند به او چهار درم زاج و دو درم اسفیداج رصاص. پس بسایند همه را نرم و بریزند برو سه وقیه آب و بگذارند پنج روز. پس بستانند از قلیه یک وقیه و از سرمه یک وقیه و از مازوی خرد کرده پنج درم و بر آتش دارند تا ثلث آب کم شود و اگر با مازو براده آهن و نشاسته باشد خوب است. پس در هم کنند آن دو آب را و بندازند قدری صمغ عربی و نشاسته و به آن بنویسند.

Ink that would appear white on black paper, and black on white paper for it is a unique ink: Acquire 4 *Deram* (60 g) pumice stone (*Qaliye*),[32] add to it 1 *Raṭl* (389.76 g) of water and shake for 7 days and what amount of water is lost, add more and shake. After the 7th day, filter it. Meanwhile gather 3 *Deram* (45 g) antimony collyrium (*Sorme-ye 'Amali*) and 3 *Deram* (45 g) beaten marcasite and a *Deram* (15 g) of litharge (Mardarsānj).

Beat each to powder separately, then add all to a mortar, and add to it 4 *Deram* (60 g) vitriol, and 2 *Deram* (30 g) white ceruse (of lead or tin) and beat to powder and bray well, then add to it 3 *Waqiye* (97.44 g) of water and allow to rest for 5 days.

Now gather from the mixtures, 1 *Waqiye* (32.48 g) pumice stone, 1 *Waqiye* (32.48 g) antimony, 5 *Deram* (75 g) beaten gall nuts and boil together until a third of the liquid is reduced. And if the gall nut water contains vitriol and starch, it shall be better. Now mix the two solutions together, then add some gum Arabic and some starch, and use as desired.

4.2.8 Golden Ink

Table 4.14 Golden ink: ingredients and measurements

Golden inks	Primary ingredients	Parts	Amount	Secondary ingredients	Parts	Amount
R1 (12th C)	Tin	–	150 g	–	–	–
	Saffron	–	45 g	–	–	–
	Gum Arabic	–	30 g	–	–	–
	Water	–	389.76 g	–	–	–
R2 (15th C)	White ceruse	–	27.84 g	–	–	–
	Vitriol	–	18.56 g	–	–	–
	Gum Arabic liquid	–	–	–	–	–
R3 (19th C)	Leaf gold	22	–	–	–	–
	Powdered gold	pieces	14.17 g	–	–	–
	Alcohol	–	–	–	–	–
	Honey	30 drops	9.5 g	–	–	–
	Gum Arabic	–	60 g	–	–	–
		–				
R4 (19th C)	Orpiment	–	36.8 g	–	–	–
	Gold leaf	45	–	–	–	–
	Blue clay	pieces	–	–	–	–
	Egg yolks	3 parts	36.8 g	–	–	–
		–				

[32] Steingass (2004, p. 987).

R1—Recipe from the treatise *Bayān al- Ṣanā 'āt* by Teflisi (12th C):

بستاند ده درم سنگ ارزیز و او را در پاتیله کند و یک رطل بکوچک آب در وی کند و ارزیز را بدو بمالد. وزان پس به آتشی نرم وی را بجوشاند تا رنگ ارزیز به آب ازو بیرون آید. پس بپالایدش و سه درم سنگ زعفران ریشه با وی بیامیزد و دیگرباره همی جوشاند چندتانکه رنگ زعفران جمله ببنداید. پس از آن آب را صافی کند و دو درم صمغ عربی کوفته بر وی ریزد. پس از آنکه سرد شود همی جنباند تا سخت نیکو و پاکیزه گردد و بکار برد.

Acquire 10 *Deram* (150 g) of tin (*Arziz*), place in a pot and add 1 *Raṭl* (389.76 g) water and mix them well. Boil it them on a steady flame until the water takes the colour of tin. Next filter and add 3 *Derams* (45 g) of saffron strands and set to boil again until saffron colour is extracted. Filter again and add 2 *Derams* (30 g) powdered gum Arabic. After it has cooled down mix it well until it gets reaches desired look and consistency and start using it.

R2—Recipe from the treatise *Resāle-ye Buqalamun* by Bahmani (15th C):

در ساختن مداد که به او بنویسند مانند طلا بستانند شش مثقال اسفیداج و چهار مثقال زاج سبز. همه در ظرف آبگینه کنند و سر او محکم بگیرند و در کوره بالائین آبگینه گران نهند یک شبانروز. پس بیرون آرند و بریزند برو قدری آب صمغ و بنویسند که عجب خوب است.

Gold coloured ink: Gather 6 *Meṯqāl* (27.84 g) white ceruse and 4 *Meṯqāl* (18.56 g) vitriol. Place in a glass container and close the lid tightly. Then place above a strong glass container and place in oven for a night and day. Take out and add to it some gum Arabic liquid and write with it for it is a unique ink.

R3—Recipe from the treatise *Kašf al-Ṣanāye* ' by 'Ali Ḥosseini (19th C)[33]

ورق طلا بیست و دو عدد، سفوف طلا نیم اونس، جوهر شراب، یعنی اسپریت سی قطره، شهد خالص سی گرین[33]، صمغ عربی چهار درم، آب باران یا آب قطره چهار اونس. اول طلا را در شهد و صمغ حل نمایند بعد از آن آب و اسپریت شامل کنند.

22 pieces of leaf gold, ½ ounce (14.17 g) powdered gold, 30 drops of alcohol (*Johar-e Sharāb*) or spirits, 30 grains (1.95 g) of honey, 4 *Derams* (60 g) gum Arabic, 4 oz of rainwater or distilled water. First mix and bray the gold in honey and gum, then add water and spirit.

R4—Recipe from the treatise *Kašf al-Ṣanāye* ' by 'Ali Ḥosseini (19th C)

مرکب طلا برای جدول کتب و غیره - زرنیخ طبقی چهار توله، ورق طلا چهل و پنج عدد، تباشیر کبود سه هسته، زرده تخم مرغ چهار توله، همه را خوب ساییده اندک آب گرم در آن ریزند و بیامیزند و ساعتی نگاه داشته که ته نشین شود و آب صاف بالایش ایستاده آهسته آهسته بریزند تا یک هفته به آب صمغ پلاس حل نموده بعد در شیشه ریخته چهار روز در آفتاب بگذارند پس از آن بکار برند.

Golden ink for framing lines of text block and other uses: 4 *Tuleh* (36.8 g) orpiment plates, 45 pieces of leaf gold, 3 parts (*Haste*) of blue clay, 4 *Tuleh* (36.8 g) egg yolks. Mix and beat well, then add some warm water, mix well, then allow for the particles to precipitate for one hour. Then pour pure clean water on top of it slowly and add filtered gum water to it over a week. Now pour in a bottle/glass container, place under sunlight for 4 days, and use.

[33] Reads: Grain. The unit of measurement equivalent to 0.065 g.

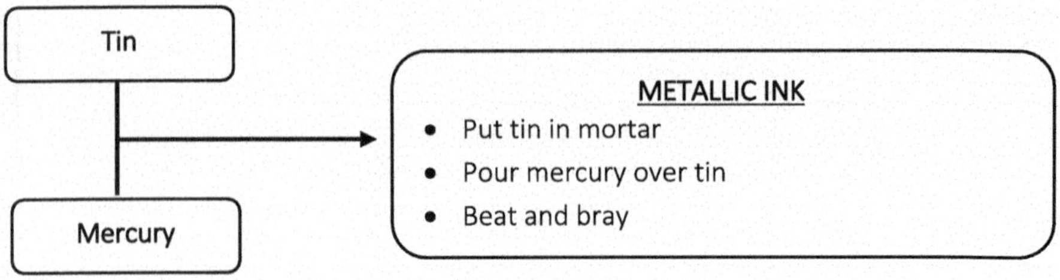

Fig. 4.11 Flowchart guide for constructing Metallic ink

Table 4.15 Metallic ink: ingredients and measurements

Metallic inks	Primary ingredients	Parts	Amount	Secondary ingredients	Parts	Amount
R1 (15th C)	Tin Quicksilver (mercury) Gum Arabic Water	1 1 – –	– – – –	– – – –	– – – –	– – – –
R2 (16th C)	Tin Quicksilver (mercury) Gum Arabic Water	1 1 – –	– – – –	– – – –	– – – –	– – – –

4.2.9 Metallic Ink

One recipe was found that mentions the use of two metals to create an ink (Fig. 4.11; Table 4.15).

R1—Recipe from the treatise *Resāle dar Bayān-e Ḵaṭ va Morakkab va Kaḡaḏ va Sāḵtan-e Ranghā* by Anonymous (15th C) (Fig 4.12):

قلعی بگیرد و سیماب، هر یکی را در برابر یکدیگر، و قلعی را بگذاد و سیماب بر وی ریزد و صلایه کند. چنان که نرم شود و آب صمغ عربی در وی ریزد و کتابت کند و بر پشت مهره زند.

> Find quicksilver (mercury) (*Simāb*) and tin (*Qal'*), in equal portions. Pour quicksilver on tin and bray well. Once softened, add liquid gum Arabic to the metal mixture and bray. Now write with it and rub it on the back of seals.

R2 Recipe from the treatise *Golestān-e Honar* by Qāżi Aḥmad Qomi (16th C):

بگیرد قلع و سیماب هریکی را برابر یکدیگر و قلعی را بگدازد و سیماب بر وی ریزد وصلایه کند چنانچه نرم شود و آب صمغ عربی در وی ریزد و کتابت کند و بر پشت مهره بمالد .

> Find quicksilver (mercury) (*Simāb*) and tin (*Qal'*), in equal portions. Pour quicksilver on tin and bray well. Once softened, add liquid gum Arabic to the metal mixture and bray. Now write with it and rub it on the back of seals.

> The exact same recipe as R1 was used by Qāżi Aḥmad a century later.

R4—Recipe from the treatise *Resāle-ye Buqalamun* by Bahmani (15th C):

در ساختن لیقة سخت زرد بستانند زرنیخ زرد تنک یک جزو و زعفران یک جزو و بسایند هر یک جدا جدا پس با هم و قدری صمغ عربی بستانند و همه را در ظرفی پاک کنند و بریزند برو آن قدر آب که صمغ او را بپوشاند و با او بنویسند.

Hard Yellow Liq: Acquire 1 part fine yellow orpiment and 1 part saffron, and grind and bray them separately, then together, then acquire gum Arabic, place on in a clean container and add enough water to cover, and write with it.

R5—Recipe from the treatise *Resāle-ye Buqalamun* by Bahmani (15th C):

در ساختن لیقه گلرنگ بستانند اسفیداج ارزیز یک جزو و سرنج یک جزو و هر دو را سحق کنند به سرکه انگوری و در دیگی کنند که به گل حکمت گرفته باشند و در کوره بالائین آبگینه گران نهند سه روز. پس بیرون آرند و برو ریزند اندکی آب مازو و چیزی از صمغ و بنویسند، والسلام.

Safflower (*Golrang*) liq ink: Gather 1 part lead ceruse (*Esfidāj-e Arziz*) and 1-part minium (*Soranj*) and make a paste with vinegar. Then place in a pot and cover with mud cement (*Gel-e Ḥekmat*), place in furnace oven, and cover with heavy glass. After three days and nights, remove, add some gall water and gum Arabic, and write with it.

4.3.7 *Golden Liqs*

Table 4.22 Golden liqs: ingredient and measurements

Golden liqs	Primary ingredients	Parts	Amount	Secondary ingredients	Parts	Amount
R1 (15th C)	Honey	1	–	–	–	–
	Talc	1	–	–	–	–
	Cyprus vitriol	1	–	–	–	–
	Rose water	1 bottle	–	–	–	–
	Gum Arabic	–	300 g	–	–	–
R2 (15th C)	Yellow vitriol	1	–	–	–	–
	Sal ammoniac	¼	–	–	–	–
R3 (15th C)	Talc	–	389.76 g	–	–	–
	Vitriol	–	250 g	–	–	–
	Vinegar	–	–	–	–	–
	Broad bean	–	–	–	–	–
	extract	–	–	–	–	–
	Saffron	–	–	–	–	–
	Gum Arabic					
R4 (15th C)	Talc	1	–	–	–	–
	Yellow	1	–	–	–	–
	orpiment	–	–	–	–	–
	Brazilwood extract					
R5 (15th C)	Poppy	–	–	–	–	–
	Myrtle extract	–	30 g	–	–	–
	Gum Arabic	–	7.5 g	–	–	–
	Water	–	–	–	–	–

R1—Recipe from the treatise *Resāle-ye Buqalamun* by Bahmani (15th C):

در ساختن لیقه طلا رنگ بستانند یک جزو عسل و یک جزو طلق و یک جزو زاج قبرسی خوب و بسایند همه را با هم خوب و بندازند در قرع و انبیق یعنی شیشه گلاب گران و بنهند بر آتش تا مصعد شود. پس بستانند آنچه مصعد شود و در ظرفی کنند و در آفتاب نهند بیست روز و در هر روز بسایند با درهمی صمغ عربی تا درو گداخته شود. آنگاه بردارند و بنویسند که خوب است.

Gold Coloured Liq: Gather 1 part honey, 1 part talc, and 1-part good quality Cyprus vitriol and grind them well together and add to a bottle of high quality rose water; set to boil till it rises. Remove the rising material, place in a bowl and leave under sunlight for 20 days and each day, add 1 *Deram* (15 g) gum Arabic until uniform consistency; write for it is a good ink.

R2—Recipe from the treatise *Resāle-ye Buqalamun* by Bahmani (15th C):

در ساختن لیقه ای دیگر بستانند یک جزو زاج زرد و هم چند چهاریک او نوشادر و بکوبند و با همدیگر بیامیزند و ببندند در پوست گاو و در تنوری نیم گرم آویزند یک شب و صباح بیرون آرند که نرم و غلیظ شده باشد. آنگاه به آن بنویسند بر جامه ها و بر پوست ها و بر هرچه خواهند.

Another Gold Liq: Beat 1-part yellow vitriol and a ¼ part sal ammoniac, then place and wrap in leather and hang in a warm oven over night. Remove in the morning at which point it should be soft and viscous. Then write with it on textile or parchment or whatever you desire.

R3—Recipe from the treatise *Resāle-ye Buqalamun* by Bahmani (15th C):

بستانند یک رطل طلق خوب و بسایند و در ظرفی کنند که چربی به آن نرسد و بندازند بروده درم توتیا و برو ریزند از سرکه صافی تیز آنقدر که او را بپوشاند و در آفتاب نهند پانجده روز. پس بردارند از آفتاب و در کیسه کنند استوار خوب و بستانند آب باقلا گرم و آن کیسه را درو فشارند و سنگریزه را درو اندازند و بمالند بسیار. آنگاه بستانند آنچه ازو بیرون می آید و داخل کنند قدری زعفران سائیده و قدری صمغ عربی. پس بنویسند به آن که مانند طلا نماید.

Gold coloured liq: Gather 1 *Raṭl* (389.76 g) Talc and grind to powder and place in a container free of oil and add to it 10 *Deram* (150 g) vitriol (*Tutiā*) and strong clear vinegar enough to cover and leave under sunlight for 15 days. Then pour into a strong muslin bag. Prepare warm broad bean (*Bāqālā*) water and place the bag in and press to extract; add some pebbles and rub well. Add to this mixture some ground saffron and gum Arabic and write with it for it shall look like gold.

R4—Recipe from the treatise *Resāle-ye Buqalamun* by Bahmani (15th C):

و چون خواهند که رنگ طلا دهد عوض زرنیخ سرخ، زرنیخ زرد کنند که به آن رنگ زرد می دهد.

Gold Liq Ink: And should you want the colour to be gold instead of red orpiment (Red Liq – R4) use yellow orpiment.

R5—Recipe from the treatise *Resāle-ye Buqalamun* by Bahmani (15th C):

در ساختن لیقة طلارنگ بستانند از شقایق آن قدر که خواهند و سیاه ازو دور کنند و سرخ او بردارند و در دیگ کنند و بر آتش نهند و بجوشـــانند تا رنگ به آب دهد. پس فرود آرند و صــاف کنند و بندازند برو از آب مورد قدر دو درم و صمغ عربی قدر او و بنویسند.

Gold Coloured Liq: Acquire desired amount poppy and clean the dark matter off, then place the red parts in a pot and place on fire to allow the colour to be extracted and mixed with water. Afterwards filter and add 2 *Dirams* (30 g) myrtle extract and a quarter of that (7.5 g) gum Arabic and write with it.

4.3.8 Silver Liqs

Table 4.23 Silver liqs: ingredient and measurements

Silver liqs	Primary ingredients	Parts	Amount	Secondary ingredients	Parts	Amount
R1 (15th C)	Talc	–	389.76 g	–	–	–
	Vitriol	–	250 g	–	–	–
	Vinegar	–	–	–	–	–
	Broad bean	–	–	–	–	–
	extract	–	–	–	–	–
	Gum Arabic					

R1—Recipe from the treatise *Resāle-ye Buqalamun* by Bahmani (15th C):

و اگر خواهی که مانند نقره نماید او را استعمال کن به غیر زعفران و با صمغ تنها که مانند نقره نماید.

Silver coloured Liq: And should want the ink (Golden liq—R3) to appear silver, do not add saffron; just the gum Arabic and it should look silver.

4.3.9 Other Liqs

Table 4.24 Other liqs: ingredient and measurements

Other liqs	Primary ingredients	Parts	Amount	Secondary ingredients	Parts	Amount
R1 (15th C)	Yellow vitriol	5	–	–	–	–
	Green vitriol	5	–	–	–	–
	Gall nut extract	–	–	–	–	–
	Saffron	–	–	–	–	–
	Gum Arabic	–	30 g	–	–	–

R1—Recipe from the treatise Resāle-ye *Buqalamun* by Bahmani (15th C):

در ساختن لیقه ای دیگر بستانند زاج زرد و زاج سبز قبرسی پنج جزو و آن قدر آب برو ریزند که او را بپوشاند و یک لحظه بگذارند تا آب ازو پاره ای قوت بستاند. آنگاه صاف کنند و نگاه دارند و بستانند زعفران خوب آن قدر که خواهند و بشویند به آب و صفت شستن او آن چنان است که بسایند و برو ریزند آب بسیار و بجنبانند و به ظرفی دیگر کنند. آنگاه بستانند آنچه به ته نشیند از غلیظ او. بسایند و در ظرفی کنند و برو ریزند آب بسیار و بجنبانند و به ظرفی کنند و به این طریق عمل مکرر به او کنند تا همه فارغ شود. پس بگذارند تا به ته نشیند و آب را بریزند و بگذارند تا خشک شود. پس بستانند آن قدر که احتیاج به او باشد و بسایند به چیزی از آب مازو که نگاه داشته بودند در اول و بندازند در آب دو درم صمغ وبسایند همه و به آن بنویسند.

Another liq ink: Acquire 5 parts of yellow vitriol and green Cyprus vitriol and cover with water (gall water) and let to set for a moment and allow the water to change and get strong by absorbing them. Then filter and keep, while desired amount of saffron is washed with water in the manner that saffron is grounded and added to a container, then water is added and shaken, and poured out several times. Finally, allow the saffron to sink to the bottom, remove the water and leave to dry. Then take desired amount of this and add to the gall water mixture you had left, add 2 *Deram* (30 g) gum Arabic, set to bray, and write with it.

4.4 Sympathetic Ink (Invisible Ink)

From all the manuscripts with recipes on ink making, the treatise *Majmu ʿ al-Ṣanāye ʿ* by anonymous (16th Century) and *Kašf al-Ṣanāye ʿ* by ʿAli Ḥosseini (19th C), a dedicated section providing guidelines on constructing inks that are invisible upon application. Most of the recipes from *Majmu ʿ al-Ṣanāye ʿ* is repeated in Ḥosseini's treatises, two centuries later (Figs. 4.13 and 4.14; Table 4.25).

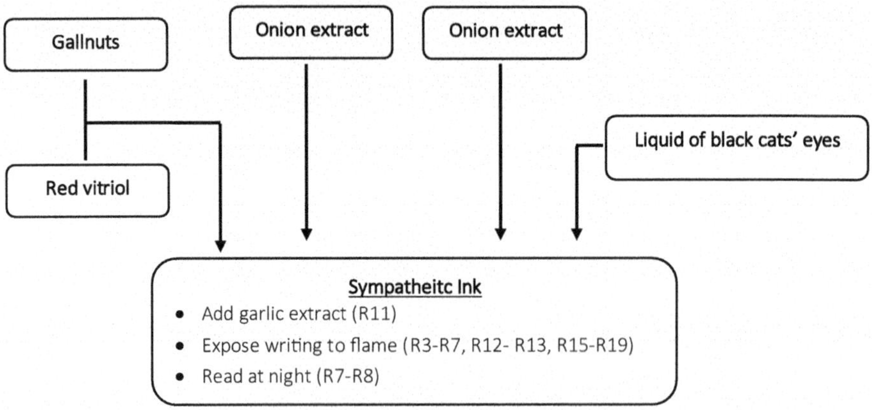

Fig. 4.13 Flowchart guide for constructing sympathetic (invisible) ink

Fig. 4.14 Recipes on making Sympathetic Inks and Other Inks from the treatise Majmu 'al-Ṣanāye' (16th C), Manuscripts 3875—Tehran University Central Library

Table 4.25 Sympathetic inks: ingredients and measurements

Sympathetic inks	Primary ingredients	Parts	Amount	Visibility aid/ reaction	Parts	Amount
R1 (12th C)	Turtle gall/alum	–	–	–	–	–
R2 (12th C)	Pigeon's blood Urine	– –	– –	– –	– –	– –
R3 (12th C)	Sal ammoniac	–	–	Heat	–	–
R4 (12th C)	Bitter orange	–	–	Heat	–	–
R5 (12th C)	Fish oil	–	–	Heat	–	–
R6 (12th C)	Onion	–	–	Heat	–	–
R7 (12th C)	Sheep's milk	–	–	Heat	–	–
R8 (12th C)	Gallnuts	–	–	Vitriol	–	–
R9 (12th C)	Gum ammoniac	–	–	Ash	–	–
R10 (12th C)	Colocynth Dioscorea	– –	– –	– –	– –	– –
R11 (16th C)	Gallnuts Red vitriol	– –	– –	Garlic extract –	– –	– –
R12 (16th C)	Onion extract	–	–	Heat	–	–

(continued)

Table 4.25 (continued)

Sympathetic inks	Primary ingredients	Parts	Amount	Visibility aid/ reaction	Parts	Amount
R13 (16th C)	Sal ammoniac	–	–	Heat	–	–
R14 (16th C)	Liquid of black cats eyes	–	–	–	–	–
R15 (19th C)	Onion extract	–	–	Heat	–	–
R16 (19th C)	Sal ammoniac	–	–	Heat	–	–
R17 (19th C)	Lime/lemon juice Onion extract	– –	– –	Heat –	– –	– –
R18 (19th C)	Liquid of black cats eyes	–	–	–	–	–
R19 (19th C)	Pigeons' blood Lamp black	– –	– –	– –	– –	– –

R1—Recipe from the treatise *Bayān al- Ṣanā'āt* by Teflisi (12th C):

اگر خواهد که نبشته کند که بشب بتوان خواند و بر وزنه، بستاند مقداری زهرهٔ کشف و بدو آنچه خواهد بنویسد. و اگر به شبّ محلول نویسد همچنین باشد.

If you want your writing to be visible at night, gather turtle gall (*Zahe-ye Kashaf*) and write with it. Writing with alum solution has similar results.

R2—Recipe from the treatise *Bayān al- Ṣanā'āt* by Teflisi (12th C):

و اگر خواهد تا نبشته کند که تا تر بود ازو بتوان خواند و چون خشک شود ناپدید گردد. بستاند خون کبوتر و او را به آب تاختن بیامیزد و بدو آنچه خواهد بنویسد.

If you want to create an invisible ink that can only be read when it is wet, and disappears when it is dry, mix pigeons blood (*K̲un-e Kabutar*) with urine, and write with it.

R3—Recipe from the treatise *Bayān al- Ṣanā'āt* by Teflisi (12th C):

و اگر خواهد که نبشته نویسد که چون ویرا بر آتش دارد سپید آید بآب نوشادر بنویسد و بنهد تا خشک شود و آن نبشته ناپدید شود و چون پیش آتش ویرا بدارد نبشته سیاه پدید آید .

If you want an invisible ink that once put on fire, it will turn white, write with sal ammoniac solution, and allow to dry. The writing will disappear and once placed above a flame, black writing will appear.

R4—Recipe from the treatise *Bayān al- Ṣanā'āt* by Teflisi (12th C):

اگر خواهد که این نبشته سرخ پدید آید بآب نارنج بنویسد و صبر کند تا خشک شود و چون برابر آتش ویرا بدارد سرخ نماید .

If you want an invisible ink to appear red, write with juice of bitter orange and allow to dry. Once placed above a flame, it will appear red in colour.

R5—Recipe from the treatise *Bayān al- Ṣanā'āt* by Teflisi (12th C):

و اگر خواهد تا زرد پدید آید روغن ماهی را سه روز بآفتاب بنهد ، وزان پس بدو بنویسد و این نبشته را بافتاب بنهد .
چون نگاه کند نبشته مانند زر پدید آید.

If you want an invisible ink to appear yellow, place fish oil under sunlight for 3 days, then write with it, and place the paper under sunlight. The writing will appear gold.

R6—Recipe from the treatise *Bayān al- Ṣanā'āt* by Teflisi (12th C):

اگر خواهد تا نبشته سبز پدید آید بآب پیاز بنویسد و بگذارد تا خشک شود. چون بر آتش بدارد سبز پدید آید.

If you want an invisible ink to appear green, write with onion juice and allow to dry. Once above a flame, the writing will appear green.

R7—Recipe from the treatise *Bayān al- Ṣanā'āt* by Teflisi (12th C):

و اگر بشیر گوسپند بنویسد چون خشک شود ناپدید گردد و چون بر آتش ویرا بدارند زرد پدید آید.

If you write with sheep milk, once dried the writing will disappear and once placed above the fire, the writing shall appear yellow.

R8—Recipe from the treatise *Bayān al- Ṣanā'āt* by Teflisi (12th C):

اگر خواهد که چیزی بنویسد که او را به آب پدید آرد باب مازو بنویسد . چون خشک شود باب زاگی بر آن نبشته دیگر
باره بنویسد سیه پدید آید.

If you want an ink that disappears, write with gallnut extract. Once dried, write over it with vitriol water and it shall appear black.

R9—Recipe from the treatise *Bayān al- Ṣanā'āt* by Teflisi (12th C):

و بدانکه اگر به آب وشه بنویسد چون خشک شود ناپدید گردد و چون خاکستر دروی بمالد نبشته سیاه پدید آید.

And know that if you write with gum ammoniac/Asafoetida (*Woshe*) and allow it to dry, it shall disappear and if you rub ash on the paper, the writing will reappear.

R10—Recipe from the treatise *Bayān al- Ṣanā'āt* by Teflisi (12th C):

و اگر خواهد که نبشته بنویسد که زود ناپدید شود ، حنظل خشک و طمیسه – و او گیاهی باشد – هر دو را بسوزاند و
باب خاکستر حنظل و معنیسای سوده بنویسد همچنان بود که نخست گفتیم تا این قدر معلوم باشد .

And if you want an ink that will disappear fast, burn dried colocynth (*Ḥanẓal-e Ḵoshk*) and Dioscorea (*Ṭamise*)—which is a plant- and write with a solution of their ash. It shall be like we first mentioned.

R11—Recipe from the treatise *Majmu' al-Ṣanāye'* by Anonymous (16th C):

اگر مازو و زاک سرخ بنویسد، پیدا نبود چون به آب سیر بنویسد خطی سیاه لطیف پیدا آید.

If a mix of gallnuts and red vitriol is used, it would be invisible but if written with garlic extract on top, a smooth black writing will appear.

R12—Recipe from the treatise *Majmu ' al-Ṣanāye* ' by Anonymous (16th C)

و اگر به آب پیاز تنها بنویسد و به آتش برند سیاه گردد.

And should you write with onion extract alone, once placed above a flame, it will turn black.

R13—Recipe from the treatise *Majmu ' al-Ṣanāye* ' by Anonymous (16th C):

و اگر به نوشادر بنویسند پیدا نبود چون بر آتش برند پیدا شود.

And should you write with sal ammoniac, it will remain invisible until the paper is placed above a flame.

R14—Recipe from the treatise *Majmu ' al-Ṣanāye* ' by Anonymous (16th C):

و اگر به آب چشم گربه سیاه بنویسند به روز پیدا نشود و به شب توان خواندن چنانکه در روز.

And should you write with the liquid of black cat's eyes, it will be invisible during the day but during the nigh time, the writing is clear as day.

R15—Recipe from the treatise *Kašf al-Ṣanāye* ' by 'Ali Ḥosseini (19th C)

اگر با آب پیاز بنویسند آن نوشته معلوم نباشد چون بر آتش گیرند سیاه درآید.

Should you write with onion extract alone, that writing will be invisible and once placed above a flame, it will turn black.

R16—Recipe from the treatise *Kašf al-Ṣanāye* ' by 'Ali Ḥosseini (19th C)

و اگر به نوشادر بنویسند پیدا نبود چون بر آتش برند پیدا شود.

And should you write with sal ammoniac, it will remain invisible until the paper is placed above a flame.

R17—Recipe from the treatise *Kašf al-Ṣanāye* ' by 'Ali Ḥosseini (19th C)

و اگر با آب لیموی تنها یا ممزوج با آب پیاز نویسند و بر آتش گیرند رنگ زرد پیدا شود.

And should you write with lime/lemon juice alone, or mixed with onion extract, once placed above flame, it will appear yellow.

R18—Recipe from the treatise *Kašf al-Ṣanāye* ' by 'Ali Ḥosseini (19th C)

اگر با آب چشم گربه سیاه نویسند روز پیدا نیست و شب معلوم خواهد شد.

Should you write with the liquid of black cat's eyes, it would be invisible during the day, and visible at night.

R19—Recipe from the treatise *Kašf al-Ṣanāye* ' by 'Ali Ḥosseini (19th C)

اگر خون کبوتر را با سیاهی مرکب بیامیزند و به آن کتابت کنند روز نتوان خواند.

If pigeons' blood is mixed with lamp black and is used, it will be unreadable during the day

4.5 Other Inks

Other types of inks are extracted from *Majmu ʿ al-Ṣanāye ʿ* by Anonymous (16th C) containing one recipe, while the rest are from ʿAli Ḥosseini's *Kašf al-Ṣanāye ʿ* (19th C). As the latter treatise uses a lot of English and Hindi terms written in Persian, I tried to decipher them as best as I could however discrepancies are to be expected (Fig. 4.15; Table 4.26).

4.5.1 *Waterproof Ink*

Table 4.26 Waterproof ink: ingredients and measurements

Waterproof Inks	Primary ingredients	Parts	Amount	Secondary ingredients	Parts	Amount
R1 (12th C)	Milk	–	–	–	–	–
	Alum	–	–	–	–	–
R2 (12th C)	Milk	–	–	–	–	–
	Lime ($Ca(OH)_2$)	–	–	–	–	–
R3 (16th C)	Breast milk	–	–	–	–	–
	Alum	–	–	–	–	–
R4 (19th C)	Breast milk	–	–	–	–	–
	Alum	–	–	–	–	–

R1—Recipe from the treatise *Bayān al- Ṣanā ʿāt* by Teflisi (12th C):

اگر خواهد تا کاغذ نبشته در آب نهد و تر نگردد و نبشته بزیان نیاید شب یمانی را خرد بساید و با شیر بیامیزد. وزآن پس هردو را به هم نیک نیک و پس ازآن از وی بر کاغذ مالد و چون خشک شود مهره اش بزند و آنچه خواهد بر وی بنویسد، چون در آب نهند کاغذ خراب نشود.

If you want the written paper to be waterproof and prevent the ink from being affected by water, grind some alum, and mix it with milk. Apply this mixture to the paper and once dried, burnish it and write on it; if this written paper is placed in water, it will not deteriorate.

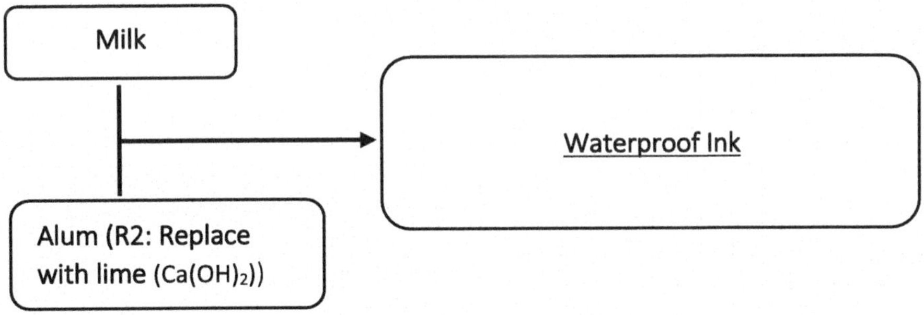

Fig. 4.15 Flowchart guide for constructing waterproof ink

R2—Recipe from the treatise *Bayān al- Ṣanā'āt* by Teflisi (12th C):

و اگر خواهد که به آب تر نشود به شیر و آهک بنویسد و رهایش کند تا خشک شود . چون ویرا در آب افکنند نبشته تر نشود

And should you want an ink that does not get wet, write with mixture of milk and lime (calcium hydroxide), and allow it to dry. If you dip the paper in water, the writing will not get wet.

R3—Recipe from the treatise *Majmu ' al-Ṣanāye '* by Anonymous (16th C):

و اگر کاغذ با شیر عورات و شب یمانی بنویسند یا بیالایند آنچه شیر رسیده بود تر نشود.

And if paper is written with a mixture of alum and breast milk, the area of application shall not be affected by water.

R4—Recipe from the treatise *Kašf al-Ṣanāye '* by 'Ali Ḥosseini (19th C)

و اگر با شیر زنان و شب یمانی بنویسند و به آب بیالایند آنچه با شیر نوشته باشند تر نشود.

And should you write with women's milk and soak paper in water, the writing shall not get wet.

4.5.2 Chinese Ink (Table 4.27)

Table 4.27 Chinese ink: ingredients and measurements

Chinese Inks	Primary ingredients	Parts	Amount	Secondary ingredients	Parts	Amount
R1 (19th C)	Lamp black Potash water Irish Moss	– – –	– – –	Musk essence Ambergris essence	– –	– –
R2 (19th C)	Lamp black Light ink Gum Arabic Indian Bael leaves Water	75 50 15 12 –	– – – – –	– – – – -	– – – – -	– – – – -

R1—Recipe from the treatise *Kašf al-Ṣanāye '* by 'Ali Ḥosseini (19th C)[36, 37, 38]

مداد چینی - دوده چراغ را در آب پطاس ملایم حل کنند چون خوب حل شود و مثل شیره گردد، اندک اندک اب پطاس بالایش بریزند و صاف نمایند، باز چند دفعه با آب شسته، خشک کنند بعد از آن دوده خشک شده را در یک ظرف نهند و جل قورباغه یا ایرش ماس[36] را در آب گرم خوب خیسانده و آمیخته که مثل شربت شود و در دوده انداخته خوب حل نمایند چند قطره آسنن اف سک[37] و اسنس آف عنبریوس[38] مخلوط کرده، بکار برند.

[36] Reads: Irish moss.

[37] Reads: Essence of 'sok'. According to Steingass (2004, p. 687) the compound word 'sokab' means musk water and hence 'sok' means musk.

[38] Reads essence of 'anbaryūs'. Probably ambergris as musk and ambergris are often used together as aromatic in ink making (Zekrgoo, 2017).

Chinese ink: Slowly mix soot of lamp with potash water until a honey like consistency, then add potash water on top little by little, and filter. Clean with water a few times and allow to dry. Then place dried soot in a plate. Now soak frog moss[39] (*Joll-e Qurbaḡe*) or Irish moss in warm water and mix till it looks like syrup/cordial, then add to the soot and bray well and add a few drops essence of musk and essence of ambergris, mix and use as desired.

A similar recipe with the name Indian ink is seen in Mitchel and Hepworth (1904: 30).

R2—Recipe from the treatise *Kašf al-Ṣanāye* ' by 'Ali Ḥosseini (19th C)

نوع دیگر- دوده هفتاد و پنج جزء، آب صمغ غلیظ پانزده جزء، مرکب کم رنگ پنجاه جزء، برگ درخت بیل دوازده جزء
با هم حل کنند به طوری که قابل نوشتن گردد، بکار برند .

Another method: 75 parts of soot, 15 parts of viscous gum Arabic liquid, 50 parts leaves of Indian Bael[40] (Bengal quince) tree. Mix and bray these ingredients well until it forms the ink, and use.

4.5.3 *Indian Ink*

Table 4.28 Indian ink: ingredients and measurements

Indian inks	Primary ingredients	Parts	Amount	Secondary ingredients	Parts	Amount
R1 (12th C)	Fir wood	–	–	–	–	–
R2 (19th C)	Lamp black Isinglass solution Gold/silver leaf –	– – – –	– – – –	Alcohol Clove oil Indian bael leaves Vinegar	– – – –	– – – –
R3 (19th C)	Lamp black *Ummu Ghailān* gum Burnt marcasite Water	– – – –	46 g 18.4 g 23 g 150 g	– – – –	– – – –	– – – –

R1—Recipe from the treatise *Bayān al- Ṣanā'āt* by Teflisi (12th C):

مداد هندی: بستاند چوب صنوبر چرب را و بر آتش نهد و کاسه رنگین زبر وی بنهد تا دودش در آنجا جمع شود. پس آنرا
در هاون کند و نیکش بساید چندانکه از سودنش عاجز شود، چون نگاه کند همچو موم نرم گشته باشد. پس دیگرباره
بمالدش تا چون مرهم شود و به سایه بنهد تا خشک شود، وزان پس به آنچه خواهد به کار برد.

[39] Steingass (2004, p. 367).

[40] Pronounced 'bīl' (بیل) in Persian. There is also a possibility that it could mean 'nīl' (نیل) (indigo) as the difference in writing is the position of a single dot.

Indian ink: Acquire oily fir wood, place in a coloured bowl and place on fire to capture and gather soot. Next place the soot in a mortar and grind well until you become exhausted and when you look at it, it has taken the form of soft wax. Bray it again till it looks like paste, then leave to dry in the shade. Use as desired.

R2—Recipe from the treatise *Kašf al-Ṣanāye* ‘ by ‘Ali Ḥosseini (19th C)[41]

مداد هندی - دوده را در آب ایزنکلاس 41حل کنند و قرصها ساخته ورق طلا و یا نقره بر آن بپیچند و بکار برند. و بدانند که اگر از سردی مداد در کاغذ سیاهی پهن شود قدري الكوهال یعنی جوهر شراب در نیم شیشه آب آمیخته به قدر نصف اونس روغن قرنفل در آن ریخته، حرکت دهند و از آن چند قطره یا زیادتر به قدر حاجت در مرکب ریزند. این آمیزش گاهی مرکب خراب نمی شود و اگر دهنیت در مرکب باشد باید برگ بیل را در قدری سرکه حل نموده در مرکب آمیزند.

Indian ink: Mix soot in isinglass solution and make small ink cakes, then wrap goldleaf or silver leaf around it, and use. And remember that if due to cold temperature, the writing on the paper bleeds, mix some alcohol in half a bottle of water, then add half ounce of clove oil and mix well. Add a few drops of this concoction to the ink. This will prevent the ink from getting damaged. And if ink contains oil, mix some leaves of Indian Bael in some vinegar, and add to ink.

R3—Recipe from the treatise *Kašf al-Ṣanāye* ‘ by ‘Ali Ḥosseini (19th C)

مرکب هندی - دوده چراغ که پاک باشد ده مثقال، صمغ ام غیلان چهار مثقال، مارقشیشای سوخته پنج مثقل، نمک ده مثقال، صبر سقوطری دو مثقال، اینها را غیر از صمغ و مارقشیشای در مسحقه سنگ سماق یا سنگی دیگر که سخت باشد اندازند و درصد درهم آب ده درهم گذارند و این را کم کم ریخته تا چهار پنج روز سحق نمایند و مارقشیشای سوخته بیامیزندران هر قدر که حل کرده اند همان قدر بسایند. این مرکب نهایت روان می شود.

و باید دانست که عمده ترکیب گرفتن دود چراغ از این قرار معین و معمول است که روغن کتان که آن را به زبان هندی السي گویند یا روغن کنجد در پیاله گلی که سابق برای چراغ معمول بوده یا چیز دیگر، پر کرده و فتیله از پارچه سفید ساخته و روشن کنند، و هرقدر فتیله پرکارتر باشد دوده زیادتر حاصل خواهد شد، روی چراغ یک پیاله بزرگ معکوسة نهند که دوده نفله نشود و چراغ هم خاموش نشود. بعد از ساعتی که دوده در قعر کاسه بسیار جمع شد، آهسته برداشته با پر مرغ یا پر کبوتر دوده را از ظرف جدا نموده در ظرف چینی یا گلی یا شیشه دهن بسته نگاهدارند که گرد وغبار به آن نرسد، بدین طریق دوده را بگیرند و اگر در میان فتیله پرهای طاوس نهاده بهم پیچیده و فتیله ساخته، مرکب لوح نما خواهد شد. و در مرکب الوان سبزگون پیدا خواهد شد. و اگر اتفاقا در دوده گل یا غبار افتد در کاسه آب پر کرده و دوده را در آب ریزند غباری که دارد ته نشین می شود و دوده خالص روی آب خواهد ماند، آن را آهسته با پر بردارند. و طریق دور کردن چربی این است که دوده را در کاغذ پیچیده و دورش خمیر گرفته مثل نان پهن کرده بر سر تنور گرم یا خاکستر گرم نهند مثل نان پخته شود، بعد گل را شکسته بیرون آرند.

10 *Meṯqāl* (46 g) pure lampblack, 4 *Meṯqāl* (18.4 g) of *Ummu Ghailān* gum (gum Arabic), 5 *Meṯqāl* (23 g) burnt marcasite, 10 *Meṯqāl* (46 g) salt, 2 *Meṯqāl* (9.2 g) aloe; aside from gum and marcasite, put the rest of these ingredients in a stone mortar. Meanwhile add the gum to 10 *Deram* (150 g) of water and add this to the mortar slowly and mix and bray for four to five days. Then add the marcasite and mix and bray again for the same amount of time; it shall be an extremely smooth ink.

And the most common way of preparing lampblack is to place linseed oil which is known as Alsi in India, or sesame oil in a clay pot which was previously used as a lamp, or something similar, fill it (with oil) and make a wick from a white cloth, and light it up. The thicker the wick, the more soot it creates. Turn

[41] Reads: Isinglass.

another pot upside down and place over the lamp in a way to capture the soot, and still allows for the oil to burn. After an hour, slowly remove the pot and collect the soot with the aid of chicken or pigeon feathers and place it in a closed lid clay, China or glass container to prevent the attraction of dirt. And should you wrap peacock feathers with the wick, the ink will be glistening (*Loḥ Namā*) and will have a green shade.

And should the lampblack contain some dirt, pour some water in a bowl, and place the lampblack on top of the water. The dirt would sink to the bottom and pure lampblack will float on top and is removed with feathers. And to remove the residing oil, place lampblack in paper and wrap it. Then secure with though and place it on an oven like you would baking bread. Then break the dough and remove the lampblack.

4.5.4 English Ink

Table 4.29 English ink: ingredients and measurements

English inks	Primary ingredients	Parts	Amount	Secondary ingredients	Parts	Amount
R1 (19th C)	Logwood	–	–	–	–	–
	Prussian blue	–	584.64 g	–	–	–
	Iron sulphate	–	170.09 g	–	–	–
	Copper acetate	–	28.34 g	–	–	–
	Sugar	–	170.09 g	–	–	–
	Gum Arabic	–	340.19 g	–	–	–
	Water	–	2 bottles	–	–	–
R2 (19th C)	Brazilwood	–	113.39 g	–	–	–
	Alum	–	–	–	–	–
	Sal ammoniac	–	–	–	–	–
	Water	–	1 bottle	–	–	–

R1—Recipe from the treatise *Kašf al-Ṣanāye* ' by 'Ali Ḥosseini (19th C)[42, 43, 44, 45]

مرکب انگریزی - لاک اود[42]، یعنی چوب بقم را در آب به قدر دو شیشه بجوشانند و آتش را بگیرند بعد یک رطل و نیم بلوکیانس[43] جوکوب، شش اونس سلفت ایرن[44]، یک اونس اسلیت اف کاپر[45]، شش اونس شکر باریک سوده و دوازده اونس صمغ ام غیلان اضافه کرده بجوشانند و بگذارند تا رنگش سیاه گردد، بعد از آن بکار برند.

Boil logwood in water, the amount of 2 bottles, then add 1 ½ *Raṭl* (584.64 g) of powdered Prussian blue (blue cyan/cyanide blue)? 6 oz (170.09 g) sulphate of iron, 1 oz (28.34 g) acetate of copper, 6 oz (170.09 g) pounded sugar, 12 oz (340.19 g) *Ummu Ghailān* gum (Gum Arabic). Boil until the liquid turns black and use as desired.

[42] Reads: Logwood.

[43] Reads: blu kiāns. Most likely cyanide blue, another name for Prussian blue which is seen repeatedly in this treatise.

[44] Reads sulphate of iron.

[45] Reads: Eslit of kāper. Most likely acetate of copper.

A similar recipe with same ingredients is found in the book Forty Centuries of Ink by Carvalho (1904, p. 196).

R2—Recipe from the treatise *Kašf al-Ṣanāye* ' by 'Ali Ḥosseini (19th C)[46], [47]

أیضاً مرکب سرخ - برازل اود[46] جوکوب چهار اونس در یک شیشه آب، روی آتش گذارده بجوشانند تا ثلث شود و شب
یمانی دو درم ساییده در آن ریخته، مخلوط کنند و بگذارند تا سرد شود پس در پارچه صاف نمایند و قدری سال امونیاک[47]
با او بیامیزند و بکار برند، بسیار خوشرنگ و روان شود.

Another method for red ink: Place 4 oz (113.39 g) pounded Brazilwood in a bottle of water and boil until third of it remains. Then add ground alum, mix, and allow to cool. Afterwards filter through cloth and add some sal ammoniac and use for it shall be a beautiful, coloured ink.

4.5.5 European Ink

Table 4.30 European ink: ingredients and measurements

European inks	Primary ingredients	Parts	Amount	Secondary ingredients	Parts	Amount
R1 (19th C)	Gall nuts Iron sulphate Gum Arabic liquid Clove Carbo-soot/ carbonic acid Water	– – – – Few drops –	389.76 g – 85.04 g – – 3.89 kg	Alcohol	– – – – – –	– – – – – –
R2 (19th C)	Gall nuts Iron sulphate Gum Arabic liquid Carbo-soot Water	– – – Five drops –	340.19 g 170.09 g 170.09 g – 8.1 kg	– – – – –	– – – – –	– – – – –

R1—Recipe from the treatise *Kašf al-Ṣanāye* ' by 'Ali Ḥosseini (19th C)[48, 49, 50, 51]

طریق ساختن مرکب اروپا - یک رطل مازو را جوکوب کرده در ده رطل آب بسیار گرم اندازند و سوم حصه آب سلفت
اف ایرن[48] در آن آمیزند و سه اونس[49] صمغ عربی در آب حل نموده و قدری قرنفل جوکوب کرده با چند قطره
کربوسوت[50] در آن اندازند یا کربانیک اسید[51] را در جوهر شراب گداخته بیامیزند و اگر مازو را ٢٤ ساعت در آب
گذارده بعد از پارچه صاف نموده اجزای منکور را در آن مخلوط نمایند بهتر از آن خواهد بود.

[46] Reads: Brāzil ud. Hence, Brazilwood.

[47] Reads: Sal ammoniac.

[48] Reads: Sulphate of iron.

[49] Reads: Ounce.

[50] Reads: Carbo-soot.

[51] Reads: Carbonic Acid (H_2CO_3).

Method of making European ink: Take a *Raṭl* (389.76 g) of gall nuts and beat them to a powder, then place them in 10 *Raṭl* (3.89 kg) of hot water. Then add iron sulphate solution followed by 3 oz (85.04 g) of liquid gum Arabic, some powdered clove (*Qaranfol*)[52] and a few drops of carbo-soot or mix carbonic acid with alcohol. And if gall nuts are left to soak for 24 h, filtered, and have the rest of the ingredients added, it shall be of better quality.

R2—Recipe from the treatise *Kašf al-Ṣanāye* ' by 'Ali Ḥosseini (19th C)

نوع دیگر- مازوی جوکوب ۱۲ اونس در پانصد توله آب سرد یک هفته بخیسانند، بعد سلفت اف ایرن شش اونس گرفته در آب ممزوج کرده در آن بیامیزند. پس شـــش اونس صـــمغ عربی را در آب حل کرده در آن بیامیزند و پنج قطره کربوسـوت در آن چکانند. و اگر خواهند که مرکب بسـیار سـیاه شـود مازو را بریان‌کرده به اسـتعمال آرند و به احتیاط بریان کنند که نه بسوزد و نه خام بماند.

Another method is to soak 12 oz (340.19 g) of powdered gall nuts in 500 *Tule* (8.1 kg) of cold water. Then dissolve 6 oz (170.09 g) of iron sulphate in water and add to the gall extract. Add 6 oz (170.09 g) gum Arabic that is dissolved in water, with five drops of carbo-soot. And if you want the ink to be extremely dark, roast the gall nuts[53] (prior to pounding) but don't burn them.

4.5.6 Gardeners (*Bāqbān*) Ink

Table 4.31 Gardeners ink: ingredients and measurements

Gardner's inks	Primary ingredients	Parts	Amount	Secondary ingredients	Parts	Amount
R1 (19th C)	Chloride of platinum	–	30 g	–	–	–
	Water	1 bottle	–	–	–	–

R1—Recipe from the treatise *Kašf al-Ṣanāye* ' by 'Ali Ḥosseini (19th C)[54, 55]

طریق مرکب باغبانان - کلورد اف پلاتینم[54] دو درم در یک باتلی[55] آب ریخته دهن آن را گرفته، نگاهدارند و از قلم که بدان نوشته می شود بر لوح چست نویسند، حروف سیاه پیدا خواهد شد که از آب و آفتاب محو نمی شود، و بدین جهت اسامی اشجار را با آن می نویسند.

Method of making Gardeners (*Bāqbān*) ink. Add 2 *Deram* (30 g) chloride of platinum in a bottle of water. Close the spout with your finger and keep. They use this ink on strong tablets/boards and it shall be a very dark ink that would not disappear due to water or sunlight; therefore, they use this ink to write family trees (*Ašjār*).

52 Steingass (2004, p. 966).

53 In my own experience, roasting otr baking the gall nuts makes it easier to pound and smash to pieces.

54 Reads: Chloride of platinum ($PtCl_2$).

55 Reads: A bottle.

Two similar recipes are found. One for writing on zinc labels by dissolving 100 g of chloride of platina in a pint of water (Beasley 1878, p. 361), and the other referred to as purple platinum marking ink, uses 1 part platinum chloride in 16 parts of water (Mitchell and Hepworth 1904, p. 201).

4.5.7 *Colour Increasing Ink*

Table 4.32 Colour increasing ink: ingredients and measurements

Colour increasing Inks	Primary ingredients	Parts	Amount	Secondary ingredients	Parts	Amount
R1 (19th C)	Gall nuts	–	56.69 g	–	–	–
	Brazilwood	–	56.69 g	–	–	–
	Sumac	–	56.69 g	–	–	–
	Protooxide of iron	–	28.34 g	–	–	–
	Water	1½ bottle	–	–	–	–
R2 (19th C)	Brazilwood	4-parts	–	–	–	–
	Gallnuts	1-part	–	–	–	–
	Sulphate protoxide of iron	1-part	–	–	–	–
	Water	12-parts	–	–	–	–

R1—Recipe from the treatise *Kašf al-Ṣanāye* ' by 'Ali Ḥosseini (19th C)[56, 57]

مرکبی که بعد از نوشتن رنگ آن افزون شود مازو دو اونس، بقم دو اونس، سوماک[56] دو انس، آب یک شیشه ونیم، بهم آمیخته و بجوشانند تا یک شیشه بماند، مصفا نموده، هراکسیس[57] یک اونس بیامیزند و زهره گاو در آن ساییده در شیشه کرده، بجنبانند و دهن شیشه را محکم نموده، در موقع حاجت بکار برند.

Ink that increases in colour once written on paper: Mix and boil 2 oz of gall (56.69 g), 2 oz (56.69 g) of Brazilwood (*Baqam*) 2 oz of Sumac (56.69 g) in 1 ½ bottle of water, until 1 bottle is left, then filter. Add and mix an ounce (28.34 g) of hydrogen peroxide, then add some ground ox gall, pour in bottle, shake, and close the bottle tightly; use as needed.

R2—Recipe from the treatise *Kašf al-Ṣanāye* ' by 'Ali Ḥosseini (19th C)[58]

أیضا بقم چهار جزء، مازو یک جزء، آب دوازده جزء، سلفت اف پروتو کسائید اف ایرن[58] یک جزء، اول هر سه را تا یک ساعت بجوشانند بعد بگذارند تا سرد شود و از پارچه بیرون نموده سلفت اف پروتوکسائید اف ایرن در او انداخته في الفور در شیشه کرده سر آن را گرفته بگذارند، وقتی حاجت بکار برند.

Another method: 4 parts Brazilwood, 1 part gall, 12 parts water, 1 part sulphate of protoxide of iron; boil the first three together for 1 h, then allow to cool, and

56 Reads: Sumac.

57 Reads: Haroxis. Most probably protooxide of iron as mentioned in the next recipe.

58 Reads: Sulphate of protoxide of iron.

filter through cloth. Then add sulphate of iron protoxide; and pour the ink in a bottle and close the lid immediately. Use as needed.

Both these appear to be versions of iron-gall ink with additional ingredients often seen in European recipes.

4.5.8 *Printing Ink*

Table 4.33 Printing ink: ingredients and measurements

Printing inks	Primary ingredients	Parts	Amount	Secondary ingredients	Parts	Amount
R1 (19th C)	Black ink Sugar candy	¾ bottle –	– 28.34 g	– –	– –	– –
R2 (19th C)	Sapanwood Garshalsu? Glycerine Chromate of potash Water	– – – – –	1.94 g 4.6 g 1.94 g 0.06 g 14.17 g	– – – – –	– – – – –	– – – – –
R3 (19th C)	Gall nuts Clove Iron sulphate Aquaforte's of sulphur Water	– – – – Print: –	127.57 g 15 g 70.87 g 52.5 kg 155.92 g 1.13 kg	– – – – –	– – – – –	– – – – -

R1—Recipe from the treatise *Kašf al-Ṣanāye* ʿ by ʿAli Ḥosseini (19th C)

مرکب چاپ - در سه ربع شیشه مرکب که رنگش سیاه باشد یک اونس نبات بیامیزند و بکار برند.

Fill ¾ of a bottle with black ink, then add and mix one ounce (28.34 g) sugar candy, and use.

R2—Recipe from the treatise *Kašf al-Ṣanāye* ʿ by ʿAli Ḥosseini (19th C)[59, 60, 61]

نوع دیگر- جوهر چوب بقم سرخ ۳۰ گرین،گرشل سو[59] ۷۱ گرین، آب خالص نیم اونس بهم آمیخته آن قدر بجوشانند که همه اجزا بهم گداخته شود پس گلسرین[60] ۳۰ گرین کرومت آف پطاس[61] یک گرین در آب گداخته، چهار گرین صمغ عربی نرم ساییده در آن بیامیزند.

Another method: Boil 30 grains (1.94 g) of sapanwood (*Baqam-e Sorḵ*) extract, 71 grains (4.6 g) of *Garšalsu*? in half ounce (14.17 g) of water and allow to mix well. Then add 30 grains (1.94 g) of glycerine and 1 grain (0.06 g) chromate of

[59] Reads: Garšalsu?

[60] Reads: Glycerine.

[61] Reads: Chromate of potash.

potash to the mixture, followed by 4 grains (0.25 g) powdered gum Arabic, and mix well.

R3—Recipe from the treatise *Kašf al-Ṣanāye* ' by 'Ali Ḥosseini (19th C)

و نوع دیگر، مازو چهار اونس و نیم قرنفل را جوکوب کرده یک درم غرابه ای که گنجایش پنجاه اونس آب داشته باشد بگیرند این اجزا را با چهل اونس آب سرد در آن بریزند و تا دو هفته بگذارند و هر روز ساعتی او را حرکت دهند که ته نشین نگردد، و بعد غرابهٔ دیگر آورده بر دهن او کاغذی که به جهت صافی مخصوص و معروف است نهاده و آن را در آن ریخته صاف کنند و هرقدر جرم آن باشد در پارچه بیفشارند و صاف کرده و در غرابه ریزند. پس یک اونس ونیم سلفت اف ایرن در آب گداخته از صافی مذکور صاف نموده در آن بیامیزند و تیزاب گوگرد خالص در پیمانه آبگینه که برای وزن عرقیات در دواخانه ها می باشد ۳۵ من نیموزن نموده در آن بیامیزند و زود بجنبانند تا اجزا مخلوط شود. ازین مرکب رنگ آسمانی نوشته می شود. اگر خواهند که برای فرود آمدن کا پی این مرکب کارآمد شود چهار اونس و نیم مازو، و از دوایی که در آغاز نسخه مرقوم است پنج اونس و نیم بگیرند و باید که از آب مقطر باشد یا از آب باران، این مداد ساخته شود.

And another method: 4.5 oz (127.57 g) of gall nuts and 1 *Deram* (15 g) powdered clove are placed in a 50-oz container, mixed, and 40 oz (1.13 kg) of cold water is added, and is left for 2 weeks while every day, it needs to be mixed for 1 h to prevent precipitation. Now bring another container, and using special filter paper, filter the content of the first container into the second. Next, 2.5 oz (70.87 g) sulphate of iron is added to water and filtered and 17.5 *Mann* (52.5 kg) pure Aquaforte's of sulphur is added after, and immediately shaken to mix all the constituents well. This will create sky blue writing ink. And for use in copy print, use 5.5 oz (155.92 g) of acid of sulphur and remember that the water must be distilled or rainwater.

4.5.9 *Baked Ink (for Seals)*

Table 4.34 Baked ink: ingredients and measurements

Baked inks	Primary ingredients	Parts	Amount	Secondary ingredients	Parts	Amount
R1 (19th C)	Indian Bael	1	–	–	–	–
	leaves	25	–	–	–	–
	Oil of lavender	100	–	–	–	–
	Lamp black	1	–	–	–	–
	Indigo					
R2 (19th C)	Asphaltum	1	–	–	–	–
	Lamp black	¼	–	–	–	–
	Spirit of	4	–	–	–	–
	turpentine	1 ½	–	–	–	–
	Linseed oil					

R1—Recipe from the treatise *Kašf al-Ṣanāye* ' by 'Ali Ḥosseini (19th C)[62]

مرکب پخته - برای ثبت نمودن مهر که از اسیت آب رنگش متغیر نشود که بیل مسحوق و بیست و پنج جزء ایل اف لوندرد[62] و صد جزء دوده، نیل یک جزء با هم گداخته و آمیخته بکار برند.

[62] Reads: Oil of lavender.

Baked Ink: For protecting seals that won't get affected by water, boil and mix pounded Indian Bael with 25 parts oil of lavender, 100 parts of soot, and 1 part indigo.

R2—Recipe from the treatise *Kašf al-Ṣanāye* ' by 'Ali Ḥosseini (19th C)[63, 64, 65, 66]

نوع دیگر، اسفالیم[63] یک جزء و ربع جزء دوده بهم مخلوط نموده مصفا نمایند، روغن السی[64] که برای مطابع درست شده باشد یک جزء و نیم در آن بیامیزند و بجوشانند به طوری که مثل شربت شکر شود اسپرت آف ترین تائین[65] چهار جزء در او بیامیزند و اگر قدری کمتر پتائین[66] در آن آمیزند بهتر است و بکار برند.

Another method: mix 1-part asphaltum, and ¼ parts of soot, add a part and a half of linseed (*Alsi*) oil used in medicine, and boil until syrup like. 4 parts spirit of turpentine is added and if a bit less of turpentine Is added, it shall be better.

A similar recipe is given by Beasley (1895:361) under the name Hauseman's ink but the recipe doesn't mention using linseed oil.

4.5.10 *Special (Maḵṣuṣ) Ink for Textiles*

Table 4.35 Special Ink for textiles: ingredients and measurements

Special inks	Primary ingredients	Parts	Amount	Secondary ingredients	Parts	Amount
R1 (19th C)	Silver chains Gum Arabic	½ container –	– 14.17 g	– –	– –	– –
R2 (19th C)	Silver nitrite Gum Arabic Water	1 0.5 6	– – –	Warm water – –	– – –	– – –

R1—Recipe from the treatise *Kašf al-Ṣanāye* ' by 'Ali Ḥosseini (19th C)[67]

مرکب مخصوص - برای نشان نمودن بر پارچه جوئین آف سیلور[67] نیم شیشه و صمغ عربی نیم اونس بهم آمیخته بکار برند.

Special ink for writing on textile: Mix half a glass container of silver chains and half ounce of gum Arabic and use as desired.

R2—Recipe from the treatise *Kašf al-Ṣanāye* ' by 'Ali Ḥosseini (19th C)[68]

أیضا ناتریت اف سیلور[68] یک جزء، آب خالص شش جزء، صمغ عربی یک جزء و نیم بهم آمیخته بگدازند، اگر غلیظ شود آب گرم اضافه نمایند.

63 Reads: Asfālim. Most likely meant to read Asphaltum.

64 Reads: Alsi. Another name for linseed or flaxseed used in Urdu and Hindi (Platts, 1997).

65 Reads: Spirit of trinitine. In actuality it is turpentine.

66 Reads: Putane. Possibly referring to turpentine.

67 Reads: ja'in a̱f silver which is most likely chain of silver.

68 Reads: Nitrite of silver.

Another method: Mix and bray a part silver nitrate with 6 parts pure water and one and half parts of gum Arabic; if it is too viscous, add some warm water.

Mitchel and Hepworth (1904, p. 200) give a recipe known as silver chloride marking ink, with similar ingredients but mention the addition of some indigo carmine.

4.5.11 Water-Floating Ink

Table 4.36 Water-floating Ink: ingredients and measurements

Water-floating inks	Primary ingredients	Parts	Amount	Secondary ingredients	Parts	Amount
R1 (19th C)	Borax	1	–	–	–	–
	Lamp black	1	–	–	–	–
	Olive/petroleum oil	2	–	–	–	–

R1—Recipe from the treatise *Kašf al-Ṣanāye* ' by 'Ali Ḥosseini (19th C)

دیگر کتابتی که بر روی آب ایستاده کنند و حروف از هم نگسلد و این کتابتی است عجیب. بیاورد بوره سرخ و به سیاهی مرکب بیامیزد و هردو را با روغن زیت یا روغن نفط تساوی کند و از آن بر روی آب کتابت نمایند، نوشته روی آب بماند.

A unique ink that stays on the surface of water and the letters remain attached: Gather red borax (*Bure*) and mix it with lamp black. Then add olive oil (*Roḡan-e Zit*) or petroleum oil (*Roḡan-e Naft*) in the equal amount of the other two. Then write on water and it shall stay on the surface.

4.5.12 Stone-Writing Ink

Table 4.37 Stone-writing ink: ingredients and measurements

Stone writing inks	Primary ingredients	Parts	Amount	Secondary ingredients	Parts	Amount
R1 (19th C)	Wax	–	–	Vinegar	–	–
	Sesame oil	–	–	–	–	–

R1—Recipe from the treatise *Kašf al-Ṣanāye* ' by 'Ali Ḥosseini (19th C)

و اگر خواهند که بر سنگ نوشته پیدا شود موم را با روغن کنجد بگدازند و هرچه خواهند بنویسند، چون سه روز بگذرد با سرکه سنگ را بشویند، نوشته پیدا شود.

If you want to write on stone, mix well wax and sesame oil and write what you please. After 3 days, wash the stone with vinegar and the writing appears.

4.5.13 Ink Removers

Table 4.38 Ink removers: ingredients and measurements

Ink removers	Primary ingredients	Parts	Amount	Secondary ingredients	Parts	Amount
R1 (12th C)	Tin ceruse Gum Arabic (liquid)	– –	– –	– –	– –	– –
R2 (12th C)	White vitriol (Alum) Pumice stone	– –	– –	– –	– –	– –
R3 (12th C)	Wax	–	–	–	–	–
R4 (12th C)	Pine gum Gum ammoniac	– –	– –	– –	– –	– –
R5 (16th C)	Tin ceruse Gum Arabic (liquid)	– –	– –	– –	– –	– –
R6 (19th C)	White vitirol Potash Vinegar	– – –	– – –	– – –	– – –	– – –
R7 (19th C)	Warm wax	–	–	–	–	–
R8 (19th C)	Sal ammoniac Borax Rat poison	1 1 1	– – –	– – –	– – –	– – –

R1—Recipe from the treatise *Bayān al- Ṣanā ʿāt* by Teflisi (12th C):

اگر خواهد که نبشته از کاغذ ناپدید کند بستاند سپیداب ارزیز و بآب صمغ عربی وی را بساید و به روی نبشتگی برانداید.
چون خشک شود مهره اش بزند.

If you want to make your writing disappear, grind some tin ceruse with some gum Arabic liquid. Apply the mixture over the writing and once dried, burnish it.

R2—Recipe from the treatise *Bayān al- Ṣanā ʿāt* by Teflisi (12th C):

و اگر زاگ سفید و ترف قلیه از هر یکی بهری بستاند و بکوبد و بسر که تر کند و بنهد تا خشک شود پس دیگرباره بکوبد
و بسر که تر کند و به روی نبشتگی بینداید آن نبشته ناپدید شود.

And if you acquire white vitriol (alum) and pumice stone (*Tarf-e Qaliye*) and beat them to powder. Slightly wet it and allow to dry. Wet it again, followed by more pounding, apply over writing and it shall disappear.

R3—Recipe from the treatise *Bayān al- Ṣanā ʿāt* by Teflisi (12th C):

و اگر پاره موم نرم کند و اندک اندک بر نبشته همی نهد و برمیدارد چند بار نوشته را بجمله پاک کند.

Slowly melt some wax, apply it to the paper, and remove the wax. If you do this several times, the writing will disappear.

R4—Recipe from the treatise *Bayān al- Ṣanā ʻāt* by Teflisi (12th C):

و اگر علک را بخساید و نبشته را بدان تر کند پس وشه را گرم کند و بروی نهد و بردارد نبشته پاک شود.

Moisten some pine gum (*'Alak*) and apply on the writing. Warm some gum ammoniac / Asafoetida (*Wushih*) and apply over the writing. Once removed, the writing will be gone.

R5—Recipe from the treatise *Golestān-e Honar* by Qāżi Aḥmad Qomi (16th C):

بستاند سفید آب ارزیز و باب صمغ آنرا بساید و بر نوشته مالد چون خشک شود مهره بزند نوشته زایل گردد.

Grind some tin ceruse with some gum Arabic liquid. Apply the mixture over the writing and leave to dry. Then burnish it and the writing will dissappear.

This recipe by Qāżi Aḥmad is almost exactly the same recipe as R1 by Teflisi (12th C).

R6—Recipe from the treatise *Kašf al-Ṣanāye* ʻ by ʻAli Ḥosseini (19th C)

و اگر زاج سفید را با قلیای سرکه سحق نمایند بعد از خشک شدن با سرکه بیامیزند بر روی حروف بمالند همه اثر حروف زایل شود و کاغذ سفید گردد.

Should you write with white vitriol and vinegar potash[69] and beat well. Afterwards mix with vinegar and apply on writing. The ink will be removed, and paper will be white.

R7—Recipe from the treatise *Kašf al-Ṣanāye* ʻ by ʻAli Ḥosseini (19th C)

و اگر موم گرم را بر روی حروف نوشته بگذارند و بردارند نیز همین عمل را کند.

Same result will be achieved if warm wax is applied on the writing.

R8—Recipe from the treatise *Kašf al-Ṣanāye* ʻ by ʻAli Ḥosseinii (19th C)[70]

و اگر نشادر و سها که[70] و سم الفار مساوی الوزن ساییده بر حروف نوشته ریزند و بر آفتاب نگاه دارند اثر حروف زایل شود.

And if sal ammoniac is mixed with borax (*Sahāke)* and rat poison (*Samm al-Fār*) are pounded and brayed in the same portions, and added to writing, once exposed to sunlight the writing will start to disappear.

[69] Reads: Potas.

[70] Reads: Sahāke which is a Hindi term for borax (abadis.ir).

Transliteration of the Persian Alphabet

Based on the Encyclopedia Iranica

Consonants		Vowels	
Persian	Transliteration	Persian	Transliteration
آ	ā	َ	a
ب	b	ِ	e
پ	p	ُ	o
ت	t	ا	ā
ث	ṯ	یٰ	ā
ج	j	ي	i
چ	č	و	u
ح	ḥ	يّ	i
خ	ḵ	وُّ	ow
د	d	وَ	aw
ذ	ḏ	یَ	ey/ay
ر	r		
ز	z		
ژ	ž		
س	s		
ش	š		
ص	ṣ		
ض	ż		
ط	ṭ		
ظ	ẓ		
ع	ʿ		
غ	ḡ		
ف	f		
ق	q		
ک	k		
گ	g		

https://doi.org/10.1007/978-3-031-52071-6

Consonants		Vowels	
Persian	Transliteration	Persian	Transliteration
ل	l		
م	m		
ن	n		
و	v/u		
ه	h		
ی ء	Y ’		

Glossary of Persian Terms

Transliteration	English	Persian
'Abir-e ašhab	Liquidamber/compound perfume	عبیر اشهب
'Āj	Ivory	عاج
'Aqiq	Agate	عقیق
'Anbar-e ašhab	Liquidamber/compound perfume	عنبر اشهب
'Araq	Essence	عرق
'Araq-e kāsni	Chicory extract	عرق کاسنی
'Araq-e nastaran	Dog rose extract	عرق نسترن
'Asal	Honey	عسل
Āb	Water	آب
Ābi	Blue	آبی
Āb-e Tāḵtan	Urine	آب تاختن
Ādāb	Manners	آداب
Āsfāltum	Asphaltum	اسفالتم
Aftimun	Epithymum (dodder of thyme)	افتیمون
Āhār	Size	آهار
Alvān	Coloured	الوان
Angur	Grape	انگور
Ašḵār	Potash	اشخار
Ātaš	Fire	آتش
Baqam	Brazilwood	بقم
Baqam-e Banafsh	Logwood	بقم بنفش
Baqam-e Sorḵ	Sappanwood	بقم سرخ
Banafsh	Purple	بنفش
Barg-e ḵār	Camel thorn	برگ خار
Berenj	Rice/brass	برنج
Bolur	Crystal	بلور
Čaḵmāq	Flint	چخماق
Čārak	750 g or 10 *Sir*	چارک

https://doi.org/10.1007/978-3-031-52071-6

Transliteration	English	Persian
Češm	Eye	چشم
Dāng	Unit of weight equivalent to 2.5 g or a 6th of anything	دانگ
Davāt	Inkwell	دوات
Deram	Unit of weight equivalent 15 g	درم
Esfidāj-e Arziz	Lead ceruse	سفیداج ارزیز
Dude	Soot/lampblack	دوده
Fariseyi	Blue-grey colour	فریسه ای
Fitileh	Wick	فتیله
Gardakān	Wallnut	گردکان
Golāb	Rose water	گلاب
Gorbe	Cat	گربه
Halile	Myrobalan	هلیله
Hāvan	Mortar	هاون
Hendi	Indian	هندی
Ḥaliyat	Treasure	حلیت
Ḥanā	Henna	حنا
Ḥanẓal	Colocynth	حنظل
Jaš	Seashell	جش
Jeld	Cover/binding	جلد
Kabud	Blue	کبود
Kafak-e Daryā	Coral	کفک دریا
Kaḡaḏ	Paper	کاغذ
Kāsni	Chicory	کاسنی
Katān	Linseed	کتان
Kāt-e Hendi	Indian Catechu	کات هندی
Ketāb	Book/manuscript	کتاب
Kochenil	Cochineal	کوچنیل
Ḵār šotor	Camel thorn	خار شتر
Ḵamir	Dough/paste	خمیر
Ḵaṭ	Script/calligraphy	خط
Ḵarboze	Rocket melon	خربزه
Ḵaṭmi	Marshmallow	خطمی
Ḵeiri-e Sorḵ	Red Mallow	خیری سرخ
Ḵun-e Kabutar	Pigeon's blood	خون کبوتر
Lājvard	Lapis lazuli	لاجورد
Lāk	Lac	لاک
Liqeh	Strings of silk inserted in inkwell	لیقه
Lo ʿāb	Mucilage	لعاب
Māhi	Fish	ماهی
Māzu	Gall nut	مازو

Transliteration	English	Persian
Mann	Unit of weight equal to 3 kg	من
Marqešišā	Marcasite	مرقشیشا
Mašq	Writing	مشق
Mes	Copper	مس
Meṣ ra '	Half a verse of poetry/half of a bait	مصرع
Meṯqāl	Unit of weight equal to 4.64 g	مثقال
Medād	Ink	مداد
Mohr	Stamp/seal	مهر
Murd	Myrtle	مورد
Morakkab	Ink	مرگب
Morakkab-e nešāstaji	Starch ink	مرگب نشاستجی
Morakkab-e ṣamḡi	Gum ink	مرگب صمغی
Mošk	Musk	مشک
Mošk -e Tabbati	Tibetan musk	مشک تبتی
Moṭavvas	Peacock coloured	مطوّس
Nārenj	Bitter orange	نارنج
Nabāt	Rock sugar	نبات
Nabāt-e Meṣri	Egyptian rock sugar/sugar candy	نبات مصری
Namak	Salt	نمک
Nastaran	Dog rose	نسترن
Nešāstaj/Nešāsteh	Starch	نشاسته/نشاستج
Ney	Reed/type of flute made of reed	نی
Ney qaṭ	A flat board used to rest the reed pen while cutting it in to shape	نی قط
Nil/Nil-e Sarāb	Indigo	نیل/نیل سراب
Noqre	Silver	نقره
Noqreyi	Silver colour	نقره ای
Noḵod	Chickpea/unit of measurement equivalent to 4.6 g	نخود
Nušādor	Sal ammoniac	نوشادر
Piyāz	Onion	پیاز
Pust	Skin/Huisk/Rind	پوست
Pust-e Kadu	Pumpkin rind	پوست کدو
Qalam	Reed pen	قلم
Qalam tarāš	Pen knife: To cut reed pens	قلم تراش
Qal '	Tin	قلع
Qaṭ	Cut	قط
Qaṭ-zan	A flat blade used to make the final cut on the tip of the reed pen	قط زن

Transliteration	English	Persian
Qebres	Cyprus	قبرس
Rang	Colour	رنگ
Rasm	Rules	رسم
Raṭl	Unit of weight equal to 389.76 g, or 84 *Meṯqāl* or12 *Uqiah*	رطل
Resāle	Treatise	رساله
Roḡan	Oil	روغن
Roḡan-e Katān	Linseed oil	روغن کتان
Roḡan-e Māhi	Fish oil	روغن ماهی
Runās	Madder	روناس
Sabz	Green	سبز
Sandarus	Sandarac	سندروس
Ṣerāṭ	Bridge	سراط
Soranj	Minium/red lead	سرنج
Sorḵ	Red	سرخ
Sorme-ye 'Amali	Antimony collyrium	سرمه عملی
Sepid/sefid	White	سپید/سفید
*Sepide-ye Ḵaye/Sefide-ye Toḵm-e Morḡ_*	Egg white	سپیده خایه/سفیده تخم مرغ
Siāh	Black	سیاه
Siāhi	Ink	سیاهی
Simāb	Quick silver—Mercury	سیماب
Sir	Unit of measurement equalt to 75 g, or 5 *Deram,* or 16 *Meṯqāl*	سیر
Seriš	A type of vegetable glue	سریش
Serišom	Glue	سریشم
Sofāl	Earthenware	سفال
Somāq	Sumac	سماق
Soṭur	Lines	سطور
Šanjarf	Cinnabar	شنجرف
Šekar	Sugar	شکر
Šir	Milk	شیر
Šaqāyeq-e Na 'mān	Red Poppy	شقایق نعمان
Ṣabr	Aloe	صبر
Ṣamḡ	Gum	صمغ
Ṣamḡ-e 'Arabi	Gum Arabic	صمغ عربی
Torki	Turkish	ترکی
Toḵm	Seed	تخم
Tule/Tulče	16.2 g or 3.5 *Meṯqāl*	توله/تولچه
Ṭalā	Gold	طلا
Ṭalāyi	Golden	طلایی
Ṭalq	Talc	طلق

Transliteration	English	Persian
Ṭavus	Peacock	طاووس
Ṭavusi	Peacock-like	طاووسی
Osul	Doctrines/rules	اصول
Uqieh	32.48 g or 7 *Meṯqāl*	اوقیه
Vasme	Woad	وسمه
Voše/Vošaq	Gum Ammoniac	وشه/وشق
Yašm	Jade	یشم
Za 'ferān	Saffron	زعفران
Zāj	Alum—Vitriol—Coperras	زاج
Zāj-e Kabud	Blue vitriol	زاج کبود
Zāj-e Qebresi	Cyprus vitriol	زاج قبرسی
Zāj-e Sabz	Green vitriol	زاج سبز
Zāj-e Sefid	White alum/Rock alum	زاج سفید
Zāj-e Sorḵ	Red vitriol	زاج سرخ
Zāj-e Torki	Turkish vitriol	زاج ترکی
Zāj-e Zard	Yellow Vitriol	زاج زرد
Zame	Rock alum	زمه
Zangār	Verdigris	زنگار
Zard	Yellow	زرد
Zard-čube	Turmeric	زردچوبه
Zarniḵ	Orpiment	زرنیخ
Zojāj	Glass	زجاج
Zahre gāv	Ox gall	زهره گاو
Ẓarf	Plate/dish	ظرف
Ẓarf-e- Sofāli	Earthen plate	ظرف سفالی

Discussion

The symbiotic relationship between ink making and calligraphy has been integral to the evolution of both crafts. Mastery in calligraphy necessitated the adept construction of inks and the crafting of reed pens by calligraphers themselves. This bespoke approach empowered calligraphers to tailor tools and materials to their individual needs and artistic preferences.

As the art of calligraphy progressed through continuous practice and experimentation, so did the formulation of inks. Calligraphers ardently sought to perfect their inks, prioritizing qualities such as permanence, water resistance, vibrant colour, and a desirable level of gloss, particularly in carbon-based inks. Preference was often given to dark/black inks, while coloured inks found application in titles and emphasis.

The transmission of knowledge within the calligraphic community occurred through master calligraphers mentoring apprentices, often their own children, and through the documentation of treatises on this art. These manuscripts served as educational resources for contemporaries and future generations, facilitating the perpetuation and evolution of calligraphy. With the migration of master calligraphers and the dissemination of knowledge through various regions, this art form reached a broader audience.

A comprehensive survey of 15 manuscripts yielded nearly 180 ink recipes, each meticulously translated from classical Persian to English. Across this diverse array of formulations, close to 70 distinct ingredients were identified, contributing to the multifaceted construction of inks. These ingredients were categorized into four primary classifications: plant-based, animal (protein) based, mineral based, and synthetic.

https://doi.org/10.1007/978-3-031-52071-6

To enhance clarity and organization, the identified ingredients were further categorized into seven distinct classes:

- Colourants: Substances responsible for imparting color to the ink.
- Binding Medium: Agents that facilitate the cohesion of ink components.
- Mordants and Additives: Elements serving to enhance ink bonding, stability and performance.
- Thickening Agents: Substances contributing to the viscosity and texture of the ink.
- Aromatics: Fragrant components incorporated into the ink.
- Preservatives: Ingredients aimed at prolonging the shelf life of the ink.
- Glossing Agents: Elements contributing to the sheen or luster of the ink.

Given the versatile nature of certain ingredients, it is important to note that an individual component might belong to more than one category, reflecting the varied and interconnected qualities inherent in the construction of these historical ink formulations.

In the realm of ink recipes, a remarkable similarity is observed across manuscripts, especially evident in carbon-tannate ink (Persian ink) recipes, where the formulation typically comprises four ingredients with consistent proportions. A commonality among most carbon-based ink recipes (carbon, Persian, Peacock, and starch ink) lies in the preparation of lamp black and the meticulous removal of excess oil. Historical texts often describe a process involving the collection of soot from burning linseed and sesame oil, encapsulating it in paper, securing it in dough, and baking it—a method aimed at eliminating residual oil through calcination. This process, described vaguely in literature, involves the conversion of some residue oil and impurities into carbon, resulting in pure carbon particles as the primary constituent of carbon ink (Mitchell and Hepworth 1904, p. 154).

Certain recipes, featuring iron sulphate or green vitriol, recommend heating the ingredient to eliminate residual "sulphur," accompanied by a noticeable change in the colour of vitriol. However, these recipes omit specific details regarding the duration of heating. The observed change in colour is likely attributed to dehydration, and prolonged heating is presumed to release sulphur dioxide and sulphur trioxide as gas, leaving behind iron oxide (FeO) (Eastaugh 2008, p. 139).

The brevity of many ink recipes notwithstanding, it is crucial to acknowledge that a considerable number of these formulations are presented in poetic form. Calligraphers, employing poetic license, have infused a creative element into their descriptions. Consequently, some statements may be interpreted as exaggerations, exemplified by claims such as the ability to write 30 sentences with a single dip of the pen into the inkwell or the notion of braying and mixing the ink for an extensive duration of 100 h. These poetic embellishments, while contributing to the aesthetic richness of the literature, warrant a discerning approach when extracting practical insights from such poetic expressions. According to Ḥujjat-Kašfi (2012, p. 21) it is impossible to write that many lines with such a small amount of ink.

Another example is the amount of time spent for mixing and braying the ink with a mortar and pestle where many recipes state a total of 100 h, or 5 days and 5 nights. Khoroush (2015) believes such sayings are used more as an emphasis on the importance of mixing the ink and that a total of 5 h of mixing will suffice.

Certain ink recipes present ingredients that, upon scrutiny, appear implausible for ink production—such as the inclusion of the liquid from black cat's eyes for creating invisible ink. These instances may be regarded as symbolic or metaphorical rather than practical instructions. Moreover, it is noteworthy that a limited number of recipes specify the quantity of water. The fluidity of the ink is thereby inferred to be contingent upon the calligrapher's personal preferences and individualized techniques, underlining the nuanced and subjective nature of ink preparation within the calligraphic tradition.

Indeed, all things considered, it is truly remarkable that master calligraphers not only meticulously documented ink recipes but also transmitted this knowledge across generations. The detailed nature of these recipes, as well as the meticulous preservation and transmission of this information, allow contemporary scholars and practitioners to recreate these inks centuries after their inception.

The depth of understanding exhibited by these calligraphers, not just in the aesthetic aspects of ink but also in their scientific applications, is a testament to their profound knowledge. Their ability to comprehend the properties and interactions of various ingredients, as well as the nuanced techniques for ink preparation, showcases a level of expertise that extends beyond artistic skill to encompass a sophisticated understanding of materials and their chemical characteristics.

The legacy left by these master calligraphers is not only a treasure trove for enthusiasts of calligraphy but also a valuable resource for researchers, conservators, and historians who seek to delve into the rich tapestry of cultural heritage and artistic practices across centuries.

Future Research

In the course of crafting this book, it became increasingly apparent that the extensive research on Persian inks, initiated in 2010, could not be adequately encapsulated within a single volume. Recognizing the depth and breadth of investigations yet to be conducted into the properties of these inks, a decision was made to expand the scope into multiple volumes.

This strategic approach not only acknowledges the richness and complexity of the ongoing research but also ensures that the exploration of Persian inks continues to unfold with precision and scholarly rigor across multiple volumes.

The forthcoming second volume, currently in progress, will be devoted to the meticulous reconstruction of inks based on the recipes elucidated in the present volume. This phase will be complemented by a comprehensive analysis of the created inks through advanced examination techniques designed to discern subtle differences among similarly coloured inks. Several techniques have already been employed in this endeavour, including water and abrasion text, cross section analysis, scanning electron microscopy (SEM) x-ray fluorescence (XRF), Fourier transform infrared (FTIR), accelerated aging, colorimetry, and a spectrum of multispectral imaging techniques:

- Visible light (VIS)
- Infrared (IR)
- Ultraviolet (UV)
- False Colour Infrared (FCIR)
- False Colour Ultraviolet (FCUV).

The outcomes of some of the examination techniques on select number of inks have been documented and are available for reference in an article authored by the researcher (Zekrgoo 2014).

 S. Zekrgoo, *Tradition and Science of Persian Ink Making*, https://doi.org/10.1007/978-3-031-52071-6

An additional area of profound interest lies in the comparative analysis of Islamic manuscripts, encompassing Persian, Arabic, Turkish, and Mughal, among others. This multifaceted investigation promises to yield invaluable insights for distinguishing inks across diverse regions, offering substantial contributions to the realms of art history, heritage studies, codicology, and conservation. The comparative study of manuscripts from different Islamic traditions holds the potential to deepen our understanding of cultural nuances, historical contexts, and the evolution of artistic practices, thereby enriching the collective knowledge base within these interdisciplinary fields.

References

Afshar I (1957) Bayān al-Ṣanā'āt. Farhang-e Irān Zamin 5:278–458

Afshar I (2011) Kāghaẕ dar zindagi va farhang-e irāni. Mirāṯ-e Maktub, Tehran

Agricola G (1950) De re metallica/Georgius Agricola; translated from the first Latin edition of 1556, with biographical introduction, annotations and appendices ... by Herbert Clark Hoover and Lou Henry Hoover. Dover Publications

Allami A (1939) The A'ini Akbari. The A'ini Akbari, The A'ini Akbari

Anonymous 16th C. Resāle dar Bayān-e Tariqe-ye Sāḵtan-e Morakkab-e Alvān va Kaḡaḏhā-ye Alvān Iran: Malek National Library

Azadi-Buyaghchi M, Afsharpur M, Haddadi M (2017) Barresi-e Ozān-e Sonnati-e Be Kār Rafte Dar Dastūrol'amalhā-ye Sāḵt-e Morakkab Dar Resālehā-ye ḵošnevisi. Faṣlnāme-Ye Ganjine-Ye Asnād 27(1):88–102

Bahesab (2015) *Vāḥed-e andāzegiri-ye vazn va jerm*. Available at: http://www.bahesab.ir/online-unit-conversion/%D9%88%D8%B2%D9%86 (Accessed: 10 Dec 2016

Barkeshli M (2016) Historical persian recipes for paper dyes. In: Restaurator international journal for the preservation of library and archival material. De Gruyter

Barkeshli M, Ataie G, Alimohammadi M, Historical analysis of materials used in Iranian paper dyeing with special reference to the effect of henna dye on paper based on scientific analysis. In: 15th Triennal Conference of ICOM-CC, New Delhi, pp 253–261

Barkeshli M, Ataie GH (2002) PH Stability of Saffron used in verdigris as an inhibitor in Persian miniature Paintings. Restaurator 23(3):154–164

Bayani M (1985) Aḥwāl va Āsār-e Khushnivisān. Elmi Publishing, Tehran

Beasley H (1878) The druggist's general receipt book : comprising a copious veterinary formulary, numerous recipes in patent and proprietary medicines, druggists' nostrums, etc.: perfumery and cosmetics, beverages, dietetic articles, and condiments, trade chemicals, scientific processes and an appendix of useful tables. Lindsay & Blakiston, Philadelphia

Bechtold T, Mussak R (2009) Handbook of natural colorants. Wiley

Blair S (2006) Islamic calligraphy. Edinburgh University Press

Botti L, Mantovani O, Ruggiero D (2005) Calcium phytate in the treatment of corrosion caused by iron gall inks: Effects on paper. Restaurator

Cardon D (2007) Natural dyes: sources, tradition, technology and science. Archetype

Carvalho DN (1904) Forty centuries of ink, or, A chronological narrative concerning ink and its backgrounds: introducing incidental observations and deductions, parallels of time and color phenomena, bibliography, chemistry, poetical effusions, citations, anecdotes and curiosa together with some evidence respecting the evanescent character of most inks of to-day and an epitome of chemico-legal ink/by David N. Carvalho. First edition. ed.: The Banks Law Publishing Co.

https://doi.org/10.1007/978-3-031-52071-6

Colbourne J (2000) A survey of methods used in the technical examination and analysis of brown inks. In: The iron gall ink meeting, Newcastle upon Tyne, 4th & 5th September 2000: postprints, pp 37–46

da Costa AC, da Fonseca NF, de Carvalho SS, dos Santos FC, Barki L, de Freitas DS, Herbst MH, Lutterbach MT (2013) Archaeometric investigations on naturally and thermally-aged iron-gall inks using different tannin sources. Cent Eur J Chem 11:1729–1739

Daniels PT (2014) The type and spread of Arabic script. The Arabic script in Africa, Brill, pp 25–39

De Michele V (1976) Minerals, their beauty and structure/Vincenzo de Michele; [line drawings by Emilio Chiusa]; with a foreword by Frank Claringbull. Orbis Publishing

Eastaugh N (2008) Pigment compendium: a dictionary and optical microscopy of historical pigments. Routledge

Encyclopedia-Britannica (2017) Ibn Muqlah. Available at: https://www.britannica.com (Accessed: 27 Nov 2017)

Ernst CW (1992) The spirit of Islamic calligraphy: Bābā Shāh Iṣfahānī's Ādāb al-mashq. J Am Orient Soc 112(2):279–286

Farhad M, Simpson MS (1993) Sources for the study of safavid painting and patronage, or Méfiez-vous de Qazi Ahmad. Muqarnas 10:286–291

Fazaeli H (1983) Atlas-e Ḵaṭ: tahqiq dar khuṭut-e islāmi. Mash'al Publications, Isfahan

Feller RL, Roy A, FitzHugh EW, Berrie BH (1993) Artists' pigments: a handbook of their history and characteristics, vol 2. National Gallery of Art, Archetype Publications, London, Washington

Feller RL, Roy A, FitzHugh EW, Berrie BH (1997) Artists' pigments: a handbook of their history and characteristics, vol 3. National Gallery of Art, Archetype Publications, London, Washington

Ghilichkhani H (2012) Adab al-Mašq. Pekareh Publications, Tehran

Hahn O (2010) Analyses of iron gall and carbon inks by means of X-ray fluorescence analysis: a non-destructive approach in the field of archaeometry and conservation science. Restaurator 3(1):41–64

Hahn O, Wilke M, Wolff T (2008) Influence of aqueous calcium phytate/calcium hydrogen carbonate treatment on the chemical composition of iron gall inks. Restaurator 29(4):235–250

Heravi MA. 16th C-a. Medād al-Ḵoṭuṭ. Ostad Reza Mayil Hirawi Personal Library, Iran

Heravi MR, 16th C-b. Ādāb al-Mašq. Malek National Library, Iran

Heravi MR 16th C-c. Rasm al-Ḵaṭ. Tehran University Central Library, Iran

Heravi MR 16th C-d. Savād al-Ḵaṭ. Malek National Library, Iran

Ḥosseini 19th C. Kašf al-Ṣanāye. Tehran University Central Library, Iran

Hosseini HMM (1997) Analysis and recognition of Persian and Arabic handwritten characters. Ph.D., University of Adelaide

Ḥujjat-Kašfi SJ (2012) Morakkab-e khushnivisi. Majma'e Ḏaḵāere Eslāmi, Qom

Iṣfahāni BŠ. 16th C. Ādāb al-Mašq. Raza Rampur Library, India

Islamic-Arts (2012) Yaqut al-Musta'simi. Available at: http://www.islamic-arts.org/Yaqut-al-Musta'simi

Jokar J (2015) Conversation with Dr. Jalil Jokar. In: Zekrgoo S (ed)

Kashani NM (2010) Resāle-ye Buqalamun Dar Ranghā. Nāme-Ye Bahārestān 11(17):17–30

Khansari AS (1973) Golestān-i Honar Ta'līf-e Qāżi Aḥmad Mir Monšī Qomī. Entešārāt-e Boniād-e Farhangi-e Irān, Tehran

Khatibi A, Sijelmassi M (1996) The splendour of Islamic calligraphy. Thames and Hudson

Kolar J, Strlič M (2004) Evaluating the effects of treatments on iron gall ink corroded documents. A new analytical methodology. Restaurator 25(2):94–103

Koroush K (2011) Conversation and workshop with Master Keikhosrow Khoroush. In: Zekrgoo S (ed)

Krekel C (1999) The chemistry of historical iron gall inks. Int J Forensic Doc Exam 5:54–58

Kuhdezi A (1831) 'Adat al-ketab'. Asiatic Society of Bengal, 635

Larcher P (2010) Pre-eslamic Aarbic—Koranic Arabic—Classical Arabic a continuum. In: Ohlig K-H, Puin G-R (eds) The hidden origins of Islam: new research into its early history. Prometheus Books, New York, pp 263–282
Maḥmudi Ā (2009) Iṣṭilāḥāt-e khošnevisi dar she'r-e shā'erān-e bozorg-e irān. Farhangestān-e Honar, Tehran
Mašhadi SA. 15th C. Ṣerāṭ al-Soṭur. Tehran University Central Library, Tehran
Mayil-Hirawi N (1993) Ketāb Ārāyi dar Tamaddon-e Eslāmi: Majmū'e-ye Rasā'el dar Zamine-ye Ḵošnevisi, Morakkab Sāzi, Kāḡaḏgari, Taḏhib va Tajlid (English title: A Collection of Articles on Penmanship, Ink Making, Papers, Gilding, and Book Binding). Mo'assese-ye Chāp va Enteshārāt-e Āstān-e Qods-e Rażavī, Mashhad
Minorsky V (1959) Calligraphers and painters: a treatise by Qadi Ahmad, Son of Mir-Munshi. Freer Gallery of Art, Washington, D.C.
Mir-Heidar H (2012a) Ma'ārifi Giyahi, vol 1. Daftare Našhre Farhangi Eslāmi, Tehran
Mir-Heidar H (2012b) Ma'ārefe Giyahi, vol 2. Daftare Našhre Farhangi Eslāmi, Tehran
Mir-Heidar H (2012c) Ma'ārefe Giyahi, vol 3. Daftare Našhre Farhangi Eslāmi, Tehran
Mir-Heidar H (2012d) Ma'ārefe Giyahi, vol 6. Daftare Našhre Farhangi Eslāmi, Tehran
Mitchell CA, Hepworth TC (1904) Inks: their composition and manufacture; including methods of examination and a full list of English patents. Charles Griffin & Company
Muslim S (2018) The book of miscellany. Available at: https://sunnah.com/riyadussaliheen/1 (Accessed: 05 July 2018
Neevel JG (1995) Phytate: a potential conservation agent for the treatment of ink corrosion caused by irongall inks. Restaurator 16(3):143–160
Neishaburi FF (2004) Barresi va Moghayeseye Dasturulamalhaye Mondarj dar Resalehaye Khushnevesieh Farsi. Nameyeh Baharestan 4(1):81–94
Phillip W (1596) A booke of secrets. Adam Islip, London
Platts JT (1997) A dictionary of Urdū, classical Hindī and English. Munshiram Manoharlal, New Delhi
Porter Y (1985) Un traité de Simi Neyšâpuri (IX e/XV e s.), artiste et polygraphe. Stud Iran 14(2):179–198
Qalam C (2009) Calligraphy qalam: an Introductin to Arabic, Ottoman and Persian calligraphy Available at: http://calligraphyqalam.com/people/mir-ali-Tabrizi.html (Accessed: 01 Nov 2016)
Qarashi MH (1996) Tarikhcheye Dayerat alMa'aref Negari dar Jahan'. Keihan Farhangi 127:22–33
Qur'an 96: 1–5
Rasekhoon (2014) Vaznhāye Marsum dar 'Aṭṭāriha va Ṭalā Forušihā. Available at: https://rasekhoon.net/article/print/914525/ (Accessed: 10 Dec 2016
Reed-College (2014) Soltān 'Ali Mašhadi: anthology of Iranian masters of calligraphy. Available at: http://www.reed.edu/persian-calligraphy/en/soltan-ali-Mašhadi/index.html (Accessed: 17 Mar 2017)
Reissland B, Gaibor AP, Ligterink F, Neevel J G (2017) Exploring the late 19th-century landscape of ink manufacturing via a collection of 90 bottles. In: ICOM-CC 18th triennial conference preprints, copenhagen, 4–8 September 2017: International Council of Museums Paris, pp 1–10
Roberts WL, Rapp GR, Weber J (1974) Encyclopedia of minerals. Van Nostrand Reinhold Company, New York
Rostamdāri HA 16th C. Ḵaṭ va Morakkab Iran: Āstān-e Qods Rażavi Library
Roxburgh DJ (2003) On the transmission and reconstruction of Arabic calligraphy: Ibn al-Bawwab and history. Stud Islam 96:39–53
Sabzevāri F. 16th C. Osul va Qavāed-e Ḵoṭuṭ-e Sette. Tehran University Central Library, Iran
Sadabadi P, Ja'fari A, Shahidabi S (2017) 'Seir-e Tahavvol-e Honar-e Khoshnevisi va 'Anva'-e Khotut dar Makatab-e Honari-ye 'Asr-e Safavi', Tarikhche-ye Khwarazmi 13:117–152
Safadi YH (1978) Islamic calligraphy. Thames and Hudson

Safwat NF, Fehérvári G, Zakariya MU (1997) The harmony of letters: Islamic calligraphy from the Tareq Rajab Museum. National Heritage Board, Singapore

Sakkal M (2016) The art of Arabic calligraphy. Available at: http://www.sakkal.com/ArtArabicCalligraphy.html (Accessed: 25 Nov 2017

Ṣayrafi A. 14th C. Ādāb-e Ḵaṭ. Parliament Library, Iran

Shahi A (2015) Khushnevisi va 'Erfān. Pekareh Publications, Tehran

Shokrollāhi E (2015) Nezhat-e Didegāni-e Sayād-e Delhā. Faslnāme-Ye Naghdi-e Ketāb 1(3):61–70

Soltany Z, Farahmand-Boroujeni H, Abed Esfehani A (2015) Justaāe-i darbāreye Morakkab-e khoshnevisi: Vajh-e Tasmiye, Āsibhā, Mavād va Raveshhāy-e Sākht. Goldaste Publications, Isfahan

Soucek PP (1979) The art of calligraphy. In: Gray B (ed) The aarts of the book in central Asia, 14th–16th centuries. UNESCO, London

Steingass F (2004) A comprehensive Persian, English dictionary: including the Arabic words and phrases to be met with in Persian literature being Johnson and Richardson's Persian, Arabic and English dictionary. Asian Educational Services. Reprint, 4, New Delhi

Storey J (1978) The thames and Hudson manual of dyes and fabrics. Thames and Hudson, London

Swider JR, Hackley VA, Winter J (2003) Characterization of Chinese ink in size and surface. J Cult Herit 4(3):175–186

Syvum (2016) Chemistry: elements and chemical symbols—Latin names. Available at: http://www.syvum.com/cgi/online/serve.cgi/squizzes/chem/periodic1b.html (Accessed: 17 July 2016)

Tan ET, Wollongong UO, Arts UOWFOC (1999) A study of kufic script in Islamic Calligraphy and its relevance to Turkish graphic art using Latin fonts in the late Twentieth Century. University of Wollongong

Thackston WM (2001) Album prefaces and other documents on the history of calligraphers and painters/Wheeler Thackston. BRILL, Leiden, Boston

Thompson JC, Lindblad CG (1996) Manuscript Inks: being a personal exploration of the materials and modes of production. The Caber Press, Portland

Ure A (1853) A dictionary of arts, manufactures, and mines: containing a clear exposition of their principles and practice. Little, Brown & Co, Boston

Wild SJ (2014) Timeline history of Henry C. Stephens Ltd, article posted on the Stephen's ink company. Available at: http://www.stephenshouseandgardens.com/assets/ugc/docs/InkCompanyTimeline (Accessed: 04 May 2022)

Zekrgoo AH. Islamic manuscripts: differentiating the sacred, religious and non-religious. In: International symposium on the care and conservation of middle eastern manuscripts. The University of Melbourne, Melbourne, pp 26–30

Zekrgoo S (2014) Methods of creating, testing and identifying traditional black persian inks. Restaurator 35(2):133–158

Zekrgoo S, Nel P, Sloggett R (2017) Peacock ink: Investigation into the constituents of the most prized ink of persia. Restaurator 38(2):205–233

SPRINGER NATURE

GPSR Compliance

The European Union's (EU) General Product Safety Regulation (GPSR) is a set of rules that requires consumer products to be safe and our obligations to ensure this.

If you have any concerns about our products, you can contact us on ProductSafety@springernature.com

In case Publisher is established outside the EU, the EU authorized representative is:

Springer Nature Customer Service Center GmbH
Europaplatz 3
69115 Heidelberg, Germany

The manufacturer's authorised representative in the EU is Springer Nature Customer Service Centre GmbH, Europaplatz 3, 69115 Heidelberg, Germany. If you have any concerns regarding our products, please contact ProductSafety@springernature.com

Printed and bound by CPI Group (UK) Ltd, Croydon, CR0 4YY
05/10/2026
02228157-0006